D0438878

SOLAR SAILS

SOLAR SAILS

A NOVEL APPROACH TO INTERPLANETARY TRAVEL

Giovanni Vulpetti
Les Johnson
Gregory L. Matloff

COPERNICUS BOOKS
An Imprint of Springer Science+Business Media

In Association with
PRAXIS PUBLISHING LTD

ISBN 978-0-387-34404-1 e-ISBN 978-0-387-68500-7
DOI 10.1007/978-0-387-68500-7

Published in the United States by Copernicus Books, an imprint of Springer Science+Business Media.

Copernicus Books
Springer Science+Business Media
233 Spring Street
New York, NY 10013
www.springer.com

Library of Congress Control Number: 2008929597

Cover Illustration: The image on the cover is a composite of NASA photographs assembled by Jordan Rosenblum

Manufactured in the United States of America.
Printed on acid-free paper.

9 8 7 6 5 4 3 2 1

Dedicated to:

My parents
Giovanni Vulpetti

Carol, my wife and companion on this life's journey
Les Johnson

My wife, partner, and colleague, C Bangs
Gregory L. Matloff

Foreword

At the time of writing, a true solar sail has yet to be flown in space. Yet despite this, there is tremendous international interest in this exciting and visionary concept. The excitement is captured in this excellent book which contains something for everyone, from a non-mathematical discussion of the principles of solar sailing to a detailed mathematical analysis of solar sail trajectories. More than that, the book places solar sailing in its proper context by providing a discussion of other propulsion technologies and highlights the benefits (and limitations) of solar sailing.

For the lay reader the book provides a complete introduction to, and discussion of, space propulsion. For the professional scientist and engineer it provides a starting point to further explore the uses of solar sailing. For all readers, it should inspire. Solar sailing is perhaps the most captivating form of spacecraft propulsion currently under development. While other advanced concepts will not make the jump from imagination to reality for many years to come, solar sailing promises to become a reality in the near term. Read this book, and then tell you friends and colleagues that some day very soon we may be literally sailing through space on a sun beam.

Colin McInnes
University of Strathclyde, Glasgow, 31 May 2007

Contents

Preface

This is one of the first books devoted to space solar sailing written in the 21st century. It is intended for both space enthusiasts (nonexperts) and those who are more technically trained. Never before has solar-sail propulsion been so close to being demonstrated via real missions around the Earth. After a number of preliminary tasks in space, the National Aeronautics and Space Administration (NASA), the European Space Agency (ESA), and the Japan Aerospace Exploration Agency (JAXA) are now designing real experimental missions to be accomplished by the first generation of solar-sail technology. Historically, we mention three serious attempts that began the solar-sail era in space. First, the solar-sail mission to the comet Halley, fostered by JPL in the 1970s, was ultimately not approved by NASA. In 1997, the precursor sailcraft Daedalus, fostered by ESA/ESTEC, received no approval from the ESA Council. In 2005, the small experimental sailcraft Cosmos-1, sponsored by the Planetary Society (U.S.), was not successful due to the failure of the Russian submarine-based launch vehicle. However, despite these aborted attempts, the problems these mission planners dealt with provided a serious base for many further studies and serious technology development activities. Strangely enough, following these "failed" attempts, theoretical research and ground demonstrations of small-sail deployment increased in number. The benefits of solar sailing are so clear and compelling that national space agencies and private organizations could not miss the chance to make a quality jump forward in space propulsion, potentially enabling exciting new science and exploration missions throughout the solar system.

This book has four parts. The first three parts are intended for the nontechnical reader who wishes to learn more about one of the most intriguing aspects of near and medium-term spaceflight: solar-sail propulsion and the missions that solar sailing will enable. These parts are completely self-consistent and self-sufficient. Various "technical boxes" have been inserted to provide the interested reader with a more technical or historical explanation. The fourth part contains the supporting mathematics, intended for more technical readers, and in particular for undergraduate students. A glossary is provided at the end of the book containing definitions of many key terms. Many topics discussed in this book are technical in nature yet the fundamental principles may be readily understood by even the most casual reader. Regardless of the reader's general interest level, the authors have made significant efforts to achieve the following goals:

- Technical correctness in all aspects of the book
- Completeness of the main topics and subtopics within the limits of a reasonably sized book
- Timeliness, as the designs, realizations, and information related to space sailing were updated up to the moment the manuscript was sent to the publisher.

Part I, Space Engines: Past and Present, contains five chapters. Chapter 1 introduces the fundamentals of spacecraft propulsion. Chapter 2 describes how rocket engines work. Chapter 3 addresses the problems and limitations of chemical, nuclear, and ion rocket propulsion. Chapter 4 considers various non-rocket technologies that may be used for space propulsion. Chapter 5 introduces the sailing concept by starting from afar—about 45 centuries ago in the Mediterranean Sea, where the Phoenicians invented a very efficient way for navigating the seas. Some of their intuitions still hold for both sailing earthly seas and in space. The authors then summarize how conventional wind sailboats work. From related physical phenomena, consider space sails—their operational analogies and their first important differences with respect to wind-powered sails. The authors subsequently introduce the amazing nature of light and its progressive scientific comprehension that began just a few centuries ago.

Part II, Space Missions by Sail, contains five chapters. Chapter 6 states that space sailing is "free," deriving propulsion from either sunlight or the solar wind. Differences between the concepts of sunlight-driven solar sails, magnetic sails, plasma sails, and electric sails are discussed. Chapter 7 is devoted to the concept of sail spacecraft, or sailcraft, and how they drive the design of a completely new class of spacecraft. Also, the concept of micro-sailcraft is introduced. Chapter 8 compares rocket propulsion and (photon) solar-sail propulsion from many practical viewpoints: design, complexity, risks, mission requirements, and range of application. Chapter 9 is devoted to exploring and developing space by sailcraft. Near-term, medium-term, long-term, and interstellar missions are discussed; sailships to other stars are given a special emphasis. Chapter 10 describes different ways of "riding" a beam of light. Sailing via laser or microwaves is discussed and compared with the so-called particle-beam sail propulsion.

Part III, Construction of Sailcraft, contains four chapters. Chapter 11 tackles the problem of designing a solar sail. There exist different sail types according to their mission aims and stabilization modes. Maneuvering a solar sail is a fundamental operation in space. This chapter explains what spacecraft attitude is and the various sail attitude control methods that may be used. Chapter 12 deals with the problem of building a sailcraft by using today's technologies or emerging technologies for tomorrow's high-performance space sailing missions. After exploring the current policies for the first solar-sail missions, the chapter introduces nanotechnology fundamentals and some of its expected features. The chapter ends by stressing what one may conceive beyond nanotechnology—a science-fiction realm indeed. Chapter 13 discusses the advancements made to date, starting from the pioneering sail/sailcraft designs and the role of the various national space agencies, and concludes with past and current private initiatives and collaborations. Chapter 14 discusses the future plans for solar sailing in the U.S., Europe, and Japan.

Part IV, Space Sailing: Some Technical Aspects, is intended for more technical readers, in particular for undergraduate students in physics, engineering, and mathematics. Although the math has been kept simple, a modest background in physics and elementary calculus is advisable. The chapters in this section contain concepts, explanations and many figures to more technically describe sailcraft missions and their feasibility. Chapter 15 is devoted to the space sources of light, and the Sun in particular. After basic optical definitions and concepts, emphasis is put on the solar electromagnetic radiation spectrum, its variability, and the measurements made in the space era by instruments on some solar-physics satellites. Total solar irradiance, a fundamental element in solar sailing, is discussed widely. Chapter 16 starts from the heliocentric and sailcraft frames of reference and shows how to get the inertial-frame thrust acceleration from the lightness vector, defined in the sailcraft frame, through momentum-transfer phenomena. The main features of the sailcraft acceleration are highlighted via reference accelerations of particular physical meaning. Chapter 17 is the central piece of Part IV. The authors show the class of sailcraft trajectories via several technical plots. Some trajectories have been designed in the past decades, some others were investigated in the first years of this century, and others have been calculated specifically for this book by means of modern (and very complex) computer codes. After a discussion of the formal sailcraft motion vector equation, the reader is introduced to general Keplerian orbits. Then, interplanetary transfer trajectories to planets are discussed. Non-Keplerian orbits are explained, as are many-body orbits and their main characteristics, and fast and very-fast solar sailing. Chapter 18 deals with the important and delicate matter of the impact of the space environment on the whole sail system design. The reader is introduced to the main environmental problems that affect a solar-sail mission, especially if it is close to the Sun.

Acknowledgments

The authors owe special thanks to their wonderful families for the comprehension, the patience, and even some very fine suggestions received in the almost two years of our efforts. It is not an easy thing to go home in the evening from our respective institutes/companies and then after a fast dinner to work on a demanding book at night and on many, many weekends.

The authors express their thanks to Dr. Salvatore Santoli for reading and commenting on Chapter 12's section on nanotechnology.

Distinguished acknowledgments go to Prof. Colin R. McInnes, who read the whole manuscript and wrote the foreword.

Particular thanks go to the publisher, Clive J. Horwood, the copy editor, and Praxis Publishing, for their precious work—expertise, suggestions to the authors, full willingness to exchanging ideas, patience, to cite just a few—that transformed very high technical areas of spaceflight into a readable book.

Giovanni Vulpetti, Les Johnson, and Greg Matloff
May 2007

Space Engines: Past and Present

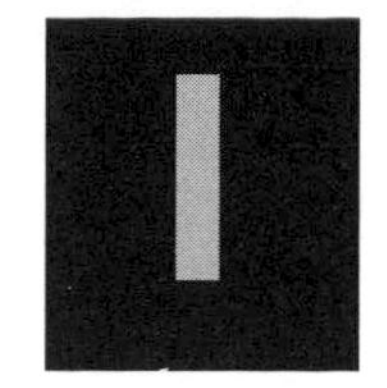

1 An Historical Introduction to Space Propulsion

We'll never know when the dream of spaceflight first appeared in human consciousness, or to whom it first appeared. Perhaps it was in the sun-baked plains of Africa or on a high mountain pass in alpine Europe. One of our nameless ancestors looked up at the night sky and wondered at the moving lights in the heavens.

Was the Moon another world similar to Earth? And what were those bright lights—the ones we call planets—that constantly change position against the background of distant stellar luminaries. Were they gods and goddesses, as suggested by the astrologers, or were they sisters to our Earth?

And if they were other worlds, could we perhaps emulate the birds, fly up to the deep heavens and visit them? Perhaps it was during a star-strewn, Moon-illuminated night by the banks of the river Nile or on the shores of the Mediterranean, as early sailing craft began to prepare for the morning trip upriver or the more hazardous sea voyage to the Cycladic Isles, that an imaginative soul, watching the pre-dawn preparations of the sailors, illuminated by those strange celestial beacons, might have wondered: If we can conquer the river and sea with our nautical technology, can we reach further? Can we visit the Moon? Can we view a planet close up?

It would be millennia before these dreams would be fulfilled. But they soon permeated the world of myth.

A Bronze-Age Astronaut

These early ponderings entered human mythology and legend. According to one Bronze-Age tale, there was a brilliant engineer and architect named Daedalus who lived on the island of Crete about 4000 years ago. For some offense, he and his son, Icarus, were imprisoned in a tower in Knossos, which was at that time the major city in Crete.

Being fed on a diet of geese and illuminating their quarters with candles, Daedalus and Icarus accumulated a large supply of feathers and wax. Being a

G. Vulpetti et al., *Solar Sails*, DOI: 10.1007/978-0-387-68500-7_1,

brilliant inventor, Daedalus fashioned two primitive hang gliders. Wings could be flapped so that the father and son could control their craft in flight.

It's not clear what their destination would be. One version of the story has the team attempting the long haul to Sicily. Another has them crossing the more reasonable 100-kilometer distance to the volcanic island of Santorini. It's interesting to note that a human-powered aircraft successfully completed the hop between Crete and Santorini only a few years ago, thereby emulating a mythological air voyage of the distant past.

Daedalus, being more mature, was cautious and content to be the first aviator. The youthful, headstrong Icarus was somewhat more ambitious. Desiring to become the first astronaut, he ignored his father's pleas and climbed higher and higher in the Mediterranean sky. Unlike modern people, the Bronze-Age Minoans had no concept of the limits of the atmosphere and the vastness of space. Icarus therefore flapped his wings, climbed higher, and finally approached the Sun. The Sun's heat melted the wax; the wings came apart. Icarus plunged to his death as his father watched in horror.

A few thousand years passed before the next fictional physical space flight was attempted. But during this time frame, several Hindu Yogi are reputed to have traveled in space by methods of astral projection.

Early Science-Fiction; The First Rocket Scientist

Starting with Pythagoras in the 6th century B.C., classical scholars began the arduous task of charting the motions of the Moon and planets, and constructing the first crude mathematical models of the cosmos. But they still had no idea that Earth's atmosphere did not pervade the universe. In what might be the first science-fiction novel, creatively entitled *True History*, the 2nd-century A.D. author Lucian of Samosata used an enormous waterspout to carry his hero to the Moon. Other authors assumed that flocks of migratory geese (this time with all their feathers firmly attached) could be induced to carry fictional heroes to the celestial realm.

What is very interesting is that all of these classical authors chose to ignore an experiment taking place during the late pre-Christian era that would pave the way to eventual cosmic travel. Hero of Alexandria, in about 50 B.C., constructed a device he called an aeolipile. Water from a boiler was allowed to vent from pipes in a suspended sphere. The hot vented steam caused the sphere to spin, in a manner not unlike a rotary lawn sprinkler. Hero did not realize what his toy would lead to, nor did the early science-fiction authors. Hero's aeolipile is the ancestor of the rocket.

Although Westerners ignored rocket technology for more than 1000 years, this was not true in the East. As early as 900 A.D., crude sky rockets were in use in China, both as weapons of war and fireworks.

Perhaps He Wanted to Meet the "Man in the Moon"

Icarus may have been the first mythological astronaut, but the first legendary rocketeer was a Chinese Mandarin named Wan Hu. Around 1000 A.D., this wealthy man began to become world-weary. He asked his loyal retainers to carry him, on his throne, to a hillside where he could watch the rising Moon. After positioning their master facing the direction of moonrise, the loyal servants attached kites and strings of their most powerful gunpowder-filled skyrockets to their master's throne.

As the Moon rose, Wan Hu gave the command. His retainers lit the fuse. They then ran for cover. Wan Hu disappeared in a titanic explosion. More than likely, his spaceflight was an elaborate and dramatic suicide. But who knows? Perhaps Wan Hu (or his fragments) did reach the upper atmosphere.

In the 13th century A.D., the Italian merchant-adventurer Marco Polo visited China. In addition to samples of pasta, the concept of the rocket returned west with him.

In post-Renaissance Europe, the imported rocket was applied as a weapon of war. It was not a very accurate weapon because the warriors did not know how to control its direction of flight. But the explosions of even misfiring rockets were terrifying to friend and foe alike.

By the 19th century, Britain's Royal Navy had a squadron of warships equipped with rocket artillery. One of these so-called "rocket ships" bombarded America's Fort McHenry during the War of 1812. Although the fort successfully resisted, the bombardment was immortalized as "the rocket's red glare" in the American national anthem, "The Star Spangled Banner."

The 19th century saw the first famous science-fiction novels. French writer Jules Gabriel Verne wrote *From the Earth to the Moon* (1865), *Twenty Thousand Leagues under the Sea* (1869), *Around the Moon* (1870), and *Around the World in Eighty Days* (1873). Particularly intriguing concepts can be found especially in the latter two books. In *Around the Moon*, Captain Nemo discovers and manages a mysterious (nonchemical) "energy", which all activities and motion of Nautilus depend on. In *Around the World in Eighty Days*, Phileas Fogg commands the crew to use his boat structure materials (mainly wood and cloth) to fuel the boat steam boiler and continue toward England. A rocket ship that (apart from its propellant) burns its useless materials progressively is an advanced concept indeed! Jules Verne is still reputed to be one of the first great originators of the science-fiction genre.

In 1902, French director Georges Méliès realized the cinematographic version of Verne's novel *From the Earth to the Moon*: in his film, *A Trip to the Moon*. Many other films describing men in space followed. For his film, Méliès invented the technique called "special effects." Thus, science-fiction cinema was born and consolidated in the first years of the 20th century, just before the terrible destruction caused by World War I.

It is surprising that science-fiction authors of the 17th to 19th centuries continued to ignore the rocket's space-travel potential, even after its military application. They employed angels, demons, flywheels, and enormous naval guns to break the bonds of Earth's gravity and carry their fictional heroes skyward. But (with the exception of Cyrano de Bergerac) they roundly ignored the pioneering efforts of the early rocket scientists.

The Dawn of the Space Age

The first person to realize the potential of the rocket for space travel was neither an established scientist nor a popular science-fiction author. He was an obscure secondary-school mathematics teacher in a rural section of Russia. Konstantin E. Tsiolkovsky (Fig. 1.1), a native of Kaluga, Russia, may have begun to ponder the physics of rocket-propelled spaceflight as early as the 1870s. He began to publish his findings in obscure Russian periodicals before the end of the 19th century. Tsiolkovsky pioneered the theory of various aspects of space travel. He considered the potential of many chemical rocket fuels, introduced the concept of the staged rocket (which allows a rocket to shed excess weight as it climbs), and was the first to investigate the notion of an orbiting space station. As will be discussed in later chapters, Tsiolkovsky was one of the first to propose solar sailing as a non-rocket form of space travel. Soviet Russia's later spaceflight triumphs

Figure 1.1. Romanian postage stamp with image of Tsiolkovsky, scanned by Ivan Kosinar. From Physics-Related Stamps Web site: www.physik.uni-frankfort.de/~jr/physstamps.html

Figure 1.2. Hermann Oberth. (Courtesy of NASA)

have a lot to do with this man. Late in his life, during the 1930s, his achievements were recognized by Soviet authorities. His public lectures inspired many young Russians to become interested in space travel. Tsiolkovsky, the recognized father of astronautics, died in Kaluga at the age of 78 on September 19, 1935. He received the last honors by state funeral from the Soviet government. In Kaluga, a museum honors his life and work.

But Tsiolkovsky's work also influenced scientists and engineers in other lands. Hermann Oberth (Fig. 1.2), a Romanian of German extraction, published his first scholarly work, *The Rocket into Interplanetary Space,* in 1923. Much to the author's surprise, this monograph became a best-seller and directly led to the formation of many national rocket societies. Before the Nazis came to power in Germany and ended the era of early German experimental cinema, Oberth created the first German space-travel special effects for the classic film *Frau Im Mund* (Woman in the Moon).

Members of the German Rocket Society naively believed that the Nazi authorities were seriously interested in space travel. By the early 1940s, former members of this idealistic organization had created the first rocket capable of reaching the fringes of outer space—the V2. With a fueled mass of about 14,000 kilograms and a height of about 15 meters, this rocket had an approximate range of 400 kilometers and could reach an altitude of about 100 kilometers. The payload of this war weapon reached its target at a supersonic speed of about 5000 kilometers per hour.

Instead of being used as a prototype interplanetary booster, the early V2s

Figure 1.3. German V2 on launch pad. (Courtesy of NASA)

(Fig. 1.3) rained down upon London, causing widespread property damage and casualties. Constructed by slave laborers in underground factories, these terror weapons had the potential to change the outcome of World War II. Fortunately, they did not.

An enlarged piloted version of the V2, called the A-10, was on the drawing boards at war's end. The A-10 could have boosted a hypersonic bomber on a trajectory that skipped across the upper atmosphere. Manhattan could have been bombed in 1946 or 1947, more than five decades before the terrorist attacks of September 11, 2001. After dropping their bombs, German skip-bomber flight crews might have turned southward toward Argentina, where they would be safely out of harm's way until the end of the war.

But America had its own rocket pioneer, who perhaps could have confronted this menace from the skies. Robert Goddard (Fig. 1.4), a physics

Figure 1.4. Robert Goddard. (Courtesy of NASA)

professor at Clark University in Massachusetts, began experimenting with liquid-fueled rockets shortly after World War I.

Goddard began his research with a 1909 study of the theory of multistage rockets. He received more than 200 patents, beginning in 1914, on many phases of rocket design and operation. He is most famous, though, for his experimental work. Funded by the Guggenheim Foundation, he established an early launch facility near Roswell, New Mexico. During the 1920s and 1940s, he conducted liquid-fueled rocket tests of increasing sophistication. One of his rockets reached the then-unheard-of height of 3000 meters! Goddard speculated about small rockets that could reach the Moon. Although he died in 1945 before his ideas could be fully realized, his practical contributions led to the development of American rocketry.

In the postwar era, the competition between the United States and the Soviet Union heated up. One early American experiment added an upper stage to a captured German V2 (Fig. 1.5). This craft reached a height of over 400 kilometers. An American-produced V2 derivative, the Viking (Fig. 1.6) was the mid-1950's precursor to the rockets that eventually carried American satellites into space.

After Russia orbited Sputnik 1 in 1957, space propulsion emerged from the back burner. Increasingly larger and more sophisticated chemical rockets were developed—first by the major space powers, and later by China, some European countries, Japan, India, and Israel. Increasingly more

Figure 1.5. A two-stage V2, launched by the United States in the postwar era. (Courtesy of NASA)

massive spacecraft, all launched by liquid or solid chemical boosters, have orbited Earth, and reached the Moon, Mars, and Venus. Robots have completed the preliminary reconnaissance of all major solar system worlds and several smaller ones. Humans have lived in space for periods longer than a year and trod the dusty paths of Luna (the Roman goddess of the Moon).

We have learned some new space propulsion techniques—low-thrust solar-electric rockets slowly accelerate robotic probes to velocities that chemical rockets are incapable of achieving. Robotic interplanetary explorers apply an elaborate form of gravitational billiards to accelerate without rockets at the expense of planets' gravitational energy. And we

Figure 1.6. A V2 derivative: the American Viking rocket. (Courtesy of NASA)

routinely make use of Earth's atmosphere and that of Mars to decelerate spacecraft from orbital or interplanetary velocities as they descend for landing.

But many of the dreams of early space travel enthusiasts remain unfulfilled. We cannot yet sail effortlessly through the void, economically tap interplanetary resources, or consider routine interplanetary transit. Human habitation only extends as far as low-Earth orbit, a few hundred kilometers above our heads, and our preliminary in-space outposts can only be maintained at great expense. And the far stars remain beyond our grasp. For humans to move further afield in the interplanetary realm as we are preparing to do in the early years of the 21st century, we need to examine alternatives to the chemical and electric rocket. The solar-photon sail—the subject of this book—is one approach that may help us realize the dream of a cosmic civilization.

Further Reading

Many sources address the prehistory and early history of space travel. Two classics are the following: Carsbie C. Adams, *Space Flight: Satellites, Spaceships, Space Stations, Space Travel Explained* (1st ed.), McGraw-Hill, New York, 1958. http://www.rarebookcellar.com/; Arthur C. Clarke, *The Promise of Space*, Harper & Row, New York, 1968.

The Minoan myth of Daedalus and Icarus is also widely available. See, for example, F. R. B. Godolphin, ed., *Great Classical Myths*, The Modern Library, New York, 1964.

Many popular periodicals routinely review space-travel progress. Two of these are the following: *Spaceflight*, published by the British Interplanetary Society; and *Ad Astra*, published by the U.S. National Space Society.

2 The Rocket: How It Works in Space

The rocket is a most remarkable device. Its early inventors could not have guessed that it would ultimately evolve into a device capable of propelling robotic and human payloads through the vacuum of space. In fact, the rocket actually works better in a vacuum than in air!

To understand rocket propulsion, we must first digress a bit into the physics of Isaac Newton.

Newtonian Mechanics and Rocket Fundamentals

A quirky and brilliant physicist, Isaac Newton framed, during the 17th century, the laws governing the motion of macroscopic objects moving at velocities, relative to the observer, well below the speed of light (300,000 kilometers per second). This discipline is called "kinematics" since it deals with motion in itself, not the causes of it. This type of physics, aptly called "Newtonian mechanics" works quite well at describing the behavior of almost all aspects common to everyday human experience, even space travel. It does not, however, accurately describe the motion of objects that are moving very fast.

To investigate kinematics of high-velocity objects moving at 20,000 kilometers per second or faster, we need to apply the results of Einstein's theory of special relativity. To consider the motion (and other properties) of microscopic objects—those much smaller than a pinhead or dust grain—we need to apply the principles of quantum mechanics. Both relativity and quantum mechanics were developed three centuries after Newton.

For macro-sized rockets moving at velocities measured in kilometers or tens of kilometers per second, newtonian physics is quite adequate. The most relevant aspects of kinematics to rocket propulsion are inertia, velocity, acceleration, and linear momentum. We will consider each of these in turn.

G. Vulpetti et al., *Solar Sails*, DOI: 10.1007/978-0-387-68500-7_2,

Inertia—Objects Resist Changes in Motion

Iron-Age scholars such as Aristotle assumed that objects move the way they do because such motion is in their nature. Although not quantifiable, such a conclusion was an improvement over the earlier Bronze-Age notion that a deity (or deities) controlled the motions of all objects.

Newton's first step in quantifying the concept of motion was to introduce the principle of inertia. All mass contains inertia—the greater the mass, the greater the inertia. Essentially, an object with mass or inertia tends to resist changes in its motion. The only way to alter the object's velocity is to act upon the object with a force. This principle is often referred to as Newton's first law; it has represented the birth of "dynamics," namely, the description of a body's motion with the inclusion of the causes that determine it.

Force and a Most Influential Equation

As a point of fact, what really separated Newton from earlier kinematicists was his elegant and most successful mathematical representation of the force concept. No longer would forces be in the province of mysterious (and perhaps) unknowable essences or natures; no longer would gods or goddesses move things at their whim. Instead, an entire technological civilization would arise based on such simple, and easily verifiable equations as Newton's relationship among force (F), mass (M), and acceleration (A).

If we are working in the MKS or standard set of units, forces are measured in units of newtons (N), masses are in kilograms (kg), and accelerations—the rate at which velocities change with time—are in meters per square second (m/s^2). The famous force equation, which is called Newton's second law, is written as follows:

$$F = MA, \tag{1}$$

or Force = Mass times Acceleration.

Let's consider what this means in practice. If a 10-newton force acts on a 1-kilogram mass, Equation 10 reveals that the force will accelerate the mass by 10 meters per square second. This force will just lift the object from the ground if it is directed upward, since Earth's gravitational acceleration (g) is 9.8 meters per square second. If the same force acts upon an object with a mass of 10 kilograms, the acceleration of the mass imparted by the force will be 1 meter per square second.

To apply Newton's second law successfully to any mode of propulsion, you must do two things. You must maximize the force and minimize the mass of the object you wish to accelerate.

Actions and Reactions

Forces, velocities, and accelerations are representatives of a type of quantity called "vectors." Unlike "scalars," which only have magnitude, vector quantities have both magnitude and direction.

We unconsciously apply the concepts of scalars and vectors all the time. Let's say that we wish to fly between London and New York. We first book a flight on an Airbus or Boeing jetliner, since such a craft can cruise at speeds of around 1000 kilometers per hour. But to minimize travel time between London and New York, we book a flight traveling in the direction of New York City—a jetliner traveling in the direction of Sydney, for example, would not do much to minimize our travel time.

Now let's examine the case of a baseball or cricket player hitting a ball with a bat. The bat is swung to impart a force on the ball, which (if all goes well from the viewpoint of the batter or bowler) flies off in the desired direction at high speed. As high-speed videotapes reveal, bats sometimes crack during the interaction. This is because a "reaction" force is imparted to the bat by the struck ball.

If you've ever fired a rifle or handgun, you've experienced action and reaction force pairs. An explosion accelerates the low-mass bullet out the gun muzzle at high speed. This is the action force. The recoil of the weapon against your shoulder—which can be painful and surprising if you are not properly braced against it—is the reaction force.

Newton's third law considers action–reaction force pairs. For every action, Newton states, there is an equal and opposite reaction. The action and reaction forces are equal in magnitude but act in opposite directions.

Jets and rockets are representative "action–reaction" propulsion systems. In a jet or chemical rocket, a controlled and contained explosion accelerates fuel to a high velocity. The ejection of this fuel from the engine nozzle is the action member of the force pair. The reaction is an equal force accelerating the engine (and structures connected to it) in the direction opposite the exhaust.

The trick with a successful jet or rocket is to minimize structural mass (and payload) and maximize fuel exhaust velocity.

Linear Momentum: A Conserved Quantity

As first-year college physics students learn, Newton's third law can be used to demonstrate that linear momentum (P) is conserved in any physical system. Linear momentum is a vector quantity, which is defined as the product of mass (M) and velocity (V) and is written $P = MV$. If the chemical reaction in the rocket's combustion chamber increases the expelled fuel's momentum by P_f, conservation of linear momentum requires that the rocket's momentum

In this text, the word *fuel* is used in a general context for simplicity. Actually, in most *chemical* rocket engines there is some substance (the proper *fuel*) that has to be burned and some other substance (the *oxidizer*) that must be present to burn the fuel. Oxidizer contains oxygen, which is required for something to burn, hence its name. (Either substance is named a *propellant*, in general.) This chemical reaction is called the *combustion*. Most of the energy released by such a reaction is found as kinetic energy of the reaction products (which are different from the propellant's molecules). They flow through a nozzle in gaseous form and achieve a final supersonic speed (the exhaust or ejection speed) with which they are exhausted away. Considered as a whole, this gas represents the reaction mass generating thrust. In solid rocket engines, fuel and oxidizer are appropriately mixed together and stored in the combustion chamber. In liquid rocket engines, fuel and oxidizer are kept separated in their tanks; they are channeled into the combustion chamber where they burn, producing the rocket's exhaust.

changes by an equal amount as that of the expelled fuel, and that this change is oppositely directed to the change in fuel momentum.

Fuel and rocket are considered as an isolated system, which is only strictly true in the depths of space. Closer to home, atmospheric air resistance tends to decrease rocket efficiency, since linear momentum of air molecules encountered by the rocket changes during the interaction. Here, the atmosphere must be considered as part of the system, which also includes rocket and fuel.

Close to a gravitating body, say near Earth's surface, a force component must always be directed upward, so the rocket can remain in flight. Even in interplanetary space, the gravitational fields of Earth, Moon, and Sun must be accounted for in estimating rocket performance.

The Rocket Equation

If one applies elementary calculus to fuel-rocket linear-momentum conservation and sets up the problem correctly, it is easy to derive the classic equation of rocket performance. We will not derive this important equation here, but will instead consider its application.

First some definitions: the mass ratio (MR) is the quotient of the total rocket mass at ignition including fuel to the mass of the vehicle when the fuel gauge is on Empty. Let's say, for example, that a particular rocket has a mass at ignition of 1 million kilograms. When the fuel has all been exhausted, the rocket's mass is 100,000 kilograms. This vehicle has a mass ratio of 1 million/ 100,000, which is exactly 10, or MR = 10.

Another significant quantity is the exhaust velocity of the rocket engine as measured by a sensor traveling with the vehicle, V_e. The final quantity expressed in the rocket equation is ΔV, which is total change of the rocket's velocity or velocity increment), measured just as all the fuel has been exhausted. All of these symbols are combined in the rocket equation as follows:

$$MR = e^{\Delta V/V_e} \qquad (2)$$

where e is approximately equal to 2.718 and is a universal constant called the "base of natural logarithms."

It is not necessary to be a rocket scientist or calculus whiz to appreciate this result. Let's say that the designers of a rocket wish the velocity increment to exactly equal the exhaust velocity. In this case, MR is 2.718 raised to the first power, or simply 2.718. For every kilogram of unfueled vehicle (payload, engines, structure, etc.), 1.718 kilograms of fuel are required.

This doesn't seem so bad, but let's examine what happens if we desire a velocity increment exactly twice the exhaust velocity. Now, MR is approximately equal to the square of 2.718, or about 7.4. For every kilogram of unfueled vehicle, 6.4 kilograms of fuel are required.

As a final illustration, consider what happens when the velocity increment is exactly three times the exhaust velocity. Now, MR becomes about 20, which means that approximately 19 kilograms of fuel are required for every kilogram of unfueled rocket.

This rapid, nonlinear increase of fuel requirement with velocity increment is called an "exponential" increase. This exponential increase demonstrates the impracticality of constructing a rocket to achieve much more than 2 or 3 times the exhaust velocity, particularly if the vehicle must overcome Earth's gravity to reach a destination in outer space.

One of the most energetic chemical rocket-fuel combination known is the liquid hydrogen/liquid oxygen combusted aboard both the American Space Shuttle and the European Ariane launchers. The highest exhaust velocity for engines of this type is about 4.5 kilometers per second.

If we desire to place a payload in low Earth orbit, say a few hundred kilometers above Earth's surface, the spacecraft must be accelerated to about 8 kilometers per second. If atmospheric drag during the early part of the rocket's climb reduces effective exhaust velocity to about 4 kilometers per second, $\Delta V/V_e$ is equal to 2. From the rocket equation, 6.4 kilograms of rocket fuel is required for every kilogram of unfueled vehicle (engines, structure, and payload). In reality, things are worse because a launcher, increasing its speed, undergoes atmospheric drag. (This drag is nothing more than friction between the rocket and the atmosphere.) The rocket's total ΔV is higher by roughly 20 to 25 percent, depending on the specific launcher design and the final orbit of payload into which it is injected.

To achieve low Earth orbit with a single-stage rocket would require advances in materials science. Strong, low-mass structures would be required for vehicle components that must withstand the high accelerations of ascent to orbit. To date, the best that has been accomplished along these lines is the American Atlas missile and space launcher of the 1960s. The Atlas had an extraordinarily thin skin. If it weren't for the pressure of the on-board fuel, the Atlas would have collapsed on the launch pad under the influence of Earth's gravity. But even using this extreme measure, the atlas was not quite a single-stage-to-orbit launcher. External boosters were used during the initial ascent phase and discarded when emptied.

If we desire a single-stage-to-orbit shuttle that is also reusable, the problem becomes even more daunting. Because of the equipment necessary to ensure reentry, the payload fraction of such craft would likely be very small, even accounting for great advances in materials and structures.

Staged Rockets

To squeeze efficiencies out of our space launchers, many of the world's space ports are located near the equator. For a west-to-east launch direction, Earth's rotation provides about 0.46 kilometers per second to the rocket, which eases the problem a bit. But geography can do little to alleviate the basic economics problem of space travel—the exhaust velocities of existing and feasible chemical launchers are simply too low!

One way around this, albeit an expensive one, is to utilize rocket stages. Basically, a big rocket lifts off from Earth's surface. Its payload consists of a smaller rocket. At burnout, the big rocket falls away and the small rocket takes over.

This approach allows us to utilize chemical rockets to achieve low Earth orbit, to escape Earth (which requires a velocity increment of about 11 kilometers per second), and to fly even faster. But there is a penalty—the payload fraction decreases dramatically as the number of stages increases and reliability issues become more pressing.

Let's consider a simple example of a 2-stage rocket. Assume that each stage has a rocket with an exhaust velocity of 4 kilometers per second and that the mass ratio of each stage is an identical 7.4. This means that at first-stage burnout, the vehicle is moving at 4 kilometers per second. At second-stage burnout, the vehicle's velocity is up to 8 kilometers per second, more than enough to achieve Earth orbit.

Next assume that the mass of the first stage is 100,000 kilograms, not including fuel, and that 20 percent of this mass is payload—the second stage in this case. The fuel required for the first stage is 620,000 kilograms

At first-stage burnout, the second stage ignites. At ignition, this stage has a mass that is 20 percent of 100,000 kilograms, or 20,000 kilograms. But to achieve the required burnout velocity, the mass ratio of the second stage is 7.4, identical to that of the first stage. At its burnout, the second stage therefore has a mass of about 2700 kilograms. If the payload fraction of the second stage is 0.2, identical to that of the first stage, about 540 kilograms of useful payload achieves Earth orbit.

Remember that the total mass of the spacecraft on the launch pad was 720,000 kilograms including fuel. Less than 0.1 percent of the on-pad vehicle mass is useful payload.

Real rockets do somewhat better, fortunately, than this simple example. The on-pad mass of Europe's Ariane 5 is about 740,000 kilograms. This launcher can inject about 10,000 kilograms into low-Earth orbit and send a bit more than half that mass toward geosynchronous orbit. But the economics are staggering—a commercial communications satellite might mass about 1 percent of the vehicle complex that propels it toward geosynchronous Earth orbit.

Chemical Rockets and Their Alternatives

The basic components of a typical chemical rocket are shown schematically in Figure 2.1. In the chemical rockets, the payload is attached usually above the fuel and oxidizer tanks. A mixture of fuel and oxidizer is delivered to the combustion chamber and then ignited in what can only be called a "controlled explosion." The product of this high-energy (exothermic) chemical reaction is squirted out the nozzle at the base of the combustion chamber as exhaust. In a reaction to the exhaust's explosive release, the rocket accelerates in the opposite direction.

In the most energetic chemical rockets, the reactants are hydrogen (H_2) fuel and oxygen (O_2), which serves as the oxidizer. For those readers a bit rusty in chemistry, the subscript "2" means that each oxygen or hydrogen molecule contains two oxygen and hydrogen atoms, respectively.

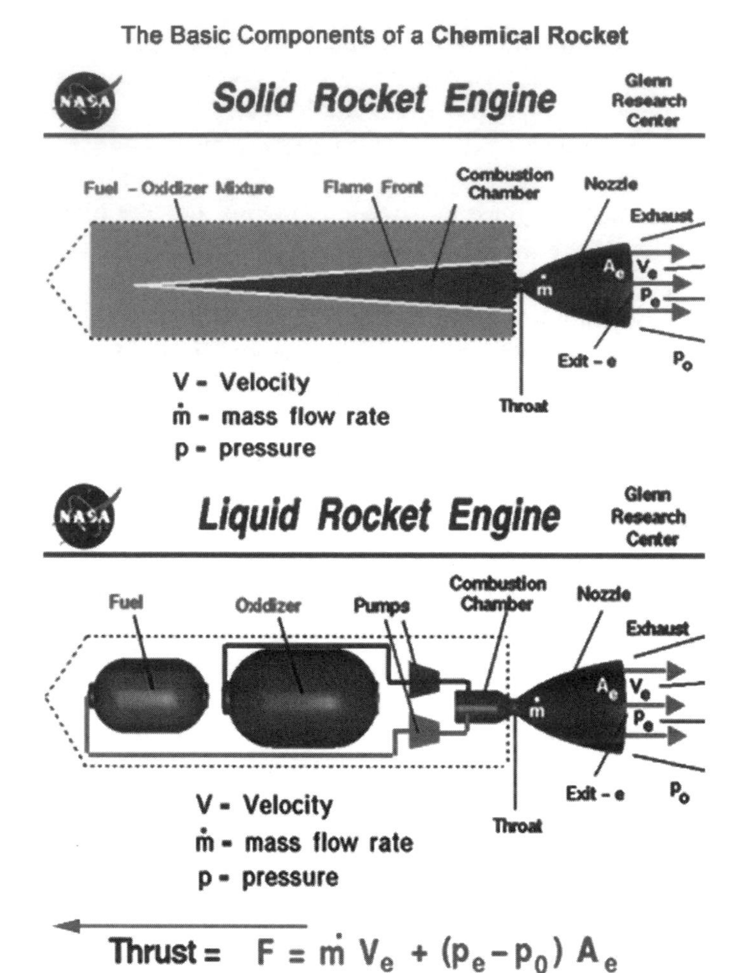

Figure 2.1. Main components of chemical engines. (Courtesy of NASA)

In many fuel/oxidizer mixtures, a device much like an auto's spark plug is required to ignite the reactants. Hydrogen and oxygen react spontaneously, however. The product of this reaction is ordinary water (H_2O) and the reaction can be expressed as follows:

$$2H_2 + O_2 \rightarrow 2H_2O \quad (3)$$

In this balanced chemical reaction, two hydrogen molecules combine with one oxygen molecule to produce two molecules of water vapor.

Some rockets use liquid fuels, such as the mixture just considered. Others, such as the space shuttle's solid boosters, burn solid fuels. There are advantages and disadvantages to both approaches.

In general, liquid fuel combinations are more energetic. But they are more difficult to store, both on Earth and in space. Many liquid rockets can be stopped and restarted. Like a skyrocket, a solid rocket once ignited burns until all fuel is exhausted.

Lots of engineering effort goes into optimizing the components shown in Figure 2.1, not to mention the complex plumbing connecting them. Engineers try to reduce the mass and the complexity of the payload faring that protects payloads as the rocket ascends through the atmosphere. Fuel tank mass is also minimized—as mentioned earlier, some fuel tanks (like those of the American Atlas boosters that orbited the Mercury astronauts) are supported by the pressure of the on-board fuel.

Combustion chambers must be low in mass, temperature resistant, and able to withstand the pressures of the expanding, ignited fuel mixtures. Millions of euros, dollars, and rubles have been expended on nozzle optimization, in an effort to squeeze the last few meters per second out of a rocket's exhaust velocity.

To overcome some of the limitations of the chemical rocket, various nonchemical rockets have been experimented with. If you don't mind a certain amount of radioactive fallout in your environment, you might consider the nuclear-thermal rocket. Ground-tested by the U.S. during the 1960s, these rockets heat a working fluid (usually water or hydrogen) to an exhaust velocity as much as twice that of the best chemical rocket. Reusable, single-stage-to-orbit nuclear-thermal shuttles are a possibility.

If you can't abide the idea of nuclear rockets streaking through the atmosphere, some of the nuclear thermal rocket's technology is applicable in the solar-thermal rocket. In this low-thrust device suited for in-space, but not ground-to-orbit, application, sunlight is focused on the working fluid, which then squirts through a nozzle at an exhaust velocity comparable to that of the nuclear-thermal rocket. (Thrust, the "action" force of the rocket, is measured in newtons and is defined as the product of the fuel flow rate in kilograms per second and the exhaust velocity in meters per second. A rocket must have a thrust greater than the rocket's weight in order to rise from the ground.)

Another low-thrust possibility is to use collected solar energy or an on-board nuclear reactor to ionize and accelerate fuel to exhaust velocities in excess of 30 kilometers per second. Several versions of these solar-electric or ion drives have seen application in robotic lunar and interplanetary missions.

From the point of view of exhaust velocity, the ultimate rocket is the nuclear-pulse drive. Nuclear-pulse rockets, which work well on paper, would be most dramatic to watch in flight since their fuel consists of nuclear charges (i.e., nuclear bombs) ignited a distance behind the craft. Fusion charges (hydrogen bombs) and even matter/antimatter combinations could conceivably propel such craft. The next chapter considers the potential and limitations of various chemical and nonchemical applications of the rocket principle.

Further Reading

Many details of chemical, electric, and nuclear rocket propulsion are reviewed in a monograph, Martin J. L. Turner, *Rocket and Spacecraft Propulsion*, 2nd ed., Springer-Praxis, Chichester, UK, 2005.

Rocket Problems and Limitations

3

Although the rockets described in the previous chapter have opened the solar system to preliminary human reconnaissance and exploration, there are severe limitations on rocket performance. This chapter focuses on these limits and what we may ultimately expect from rocket-propelled space travel.

Limits of the Chemical Rocket

A common science-fiction theme during the 1950s was the exploration of the Moon by single-stage, reusable chemical rockets. Sadly, this has not come to pass. And because of the fact that the exhaust velocities of even the best chemical rockets may never exceed 5 kilometers per second, this dream may always remain within the realm of fantasy.

During the late 1960s and early 1970s, the United States launched 9 crews of three astronauts each to lunar orbit or the Moon's surface. An appreciation of the chemical rocket's severe limitations for large-scale application beyond low Earth orbit can be arrived at by consideration of these NASA Apollo expeditions.

Everything about Apollo's Saturn V booster is gargantuan. Standing on its launch pad, this craft was 110.6 meters high, taller than the Statue of Liberty. It had a fully fueled prelaunch mass of about 3 million kilograms. Of this enormous mass, only 118,000 kilograms reached low Earth orbit and 47,000 kilograms departed on a translunar trajectory. But the Apollo command modules that safely returned the three-astronaut crews and their cargoes of Moon rocks to Earth had heights of only 3.66 meters and diameters of 3.9 meters.

The Apollo lunar expeditions were a splendid human and technological achievement. But they did not lead to the economic development or settlement of the Moon. In fact, the economics of lunar travel using chemical rocketry has been compared with a European traveler who wishes to visit the U.S. Being exceptionally wealthy, she commissions the construction of her

G. Vulpetti et al., *Solar Sails*, DOI: 10.1007/978-0-387-68500-7_3,

own private, full-scale Airbus, for an investment of a billion euros or so. She flies the aircraft to New York, parachutes out above the Empire State Building, and allows the entire aircraft to plunge into the Atlantic Ocean. You could not afford a great many intercontinental visits if that were the only way to go!

> By pushing chemical-rocket technology and materials science to their limits (perhaps in commercial efforts directed by those promoting space tourism), we may ultimately produce a reusable two-stage or even single-stage Earth-to-orbit shuttlecraft. But payload will be limited. Orbital construction will be required if we wish to venture further afield in the cosmic realm. Chemical rocket costs will severely limit the number of lunar and interplanetary missions fielded by even the wealthiest nations.

Nuclear and Solar Thermal Rockets: An Improvement with Issues

Let's look at various nonchemical rocket approaches in an attempt to overcome some of these limitations. Two options are the nuclear-thermal or solar-thermal rocket, in which the energy output of a nuclear reactor or solar collector is used to heat a working fluid (e.g., hydrogen) to a high exhaust velocity (Fig. 3.1). If the working fluid is hydrogen, exhaust velocities of 8 to 10 kilometers per second are possible, about twice those of the best-performing chemical rockets.

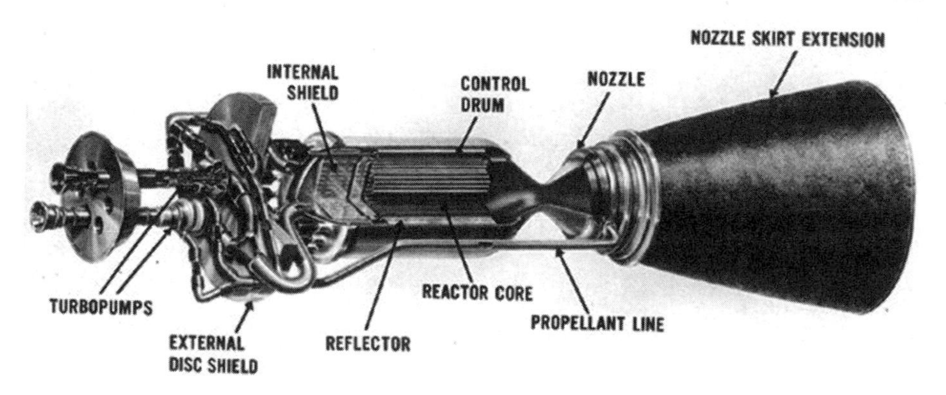

Figure 3.1. The NASA NERVA nuclear-thermal rocket concept. (Courtesy of NASA)

Figure 3.2. The NASA KIWI nuclear-thermal rocket reactor on its test stand. (Courtesy of NASA)

During the 1960s, nuclear-thermal rockets such as NASA's KIWI (Fig. 3.2) were subject to elaborate ground tests. They are high-thrust devices and are at least as reliable as their chemical brethren. Why haven't we seen the emergence of single-stage-to-orbit nuclear-thermal shuttles?

One issue with this technology is environmental pollution. Because of mass limitations, no ground-launched economical nuclear rocket could be completely shielded. As a point of fact, a lot of additional mass has to be employed for blocking all nuclear radiations. Invariably, some radioactive fallout will escape to the atmosphere.

Another problem is nuclear proliferation. If many governmental and private space agencies began to employ this technology for dozens of launches per year, what type of security measures might be required to protect the nuclear fuel from terrorists and agents of rogue states?

It would be possible to launch the reactor in a safe, inert mode, and turn it on well above Earth's atmosphere. Although this pollution-free option will do little to reduce launch costs, it might have the potential to improve the economics of lunar and interplanetary travel.

There are two problems with this approach. First and foremost is the difficulty of storing the required hydrogen for long durations in the space environment. This low-molecular-mass gas tends to evaporate rapidly into the space environment unless elaborate (and massive) precautions are taken. Nuclear rocket designers could switch to fuels other than hydrogen. But exhaust velocity decreases with increasing fuel molecular mass, and the advantage of nuclear over chemical would soon vanish.

A second problem involves nuclear-fission-reactor technology. While it is certainly possible to launch an inert reactor toward space to minimize radioactive pollution from a catastrophic launch accident, it is not possible to turn the reactor off completely once fission has been initiated. A nuclear-thermal-propelled interplanetary mission would have to contend with the problem of disposing spent nuclear stages in safe solar orbits.

The solar-thermal rocket replaces the reactor with a solar concentrator such as a thin-film Fresnel lens. Although exhaust velocities for solar-thermal rockets fueled with molecular hydrogen are comparable to those of nuclear-thermal hydrogen rockets, the diffuse nature of solar energy renders them low-thrust devices. No solar-thermal rocket will ever lift itself off the ground. Typical accelerations for these devices, in fact, are of the order of 0.01 Earth surface gravities. Major applications of this technology might be for orbital transfer—like the economical delivery of communications satellites to geosynchronous Earth orbit.

One should note that, strictly speaking, a solar-powered rocket is not a rocket because the energy for heating the propellant does not reside in the vehicle. However, such energy is always much, much less than the propellant mass times c^2, the square of the speed of light in vacuum. Thus, for the space flights we are considering here, we can continue to consider it as a rocket.

Solar and Nuclear Electric Rockets—The Ion Drive

Another nonchemical rocket option is the so-called electric rocket or ion drive. In the electric rocket (Fig. 3.3), sunlight or nuclear energy is first used to ionize fuel into electrons with negative electric charges and ions with positive electric charges. Solar- or nuclear-derived electricity is then directed to electric thrusters, which are utilized to accelerate fuel ions (and electrons) to exhaust velocities of 30 kilometers per second or higher (Fig. 3.4). Typical accelerations from these low-thrust devices are 0.0001

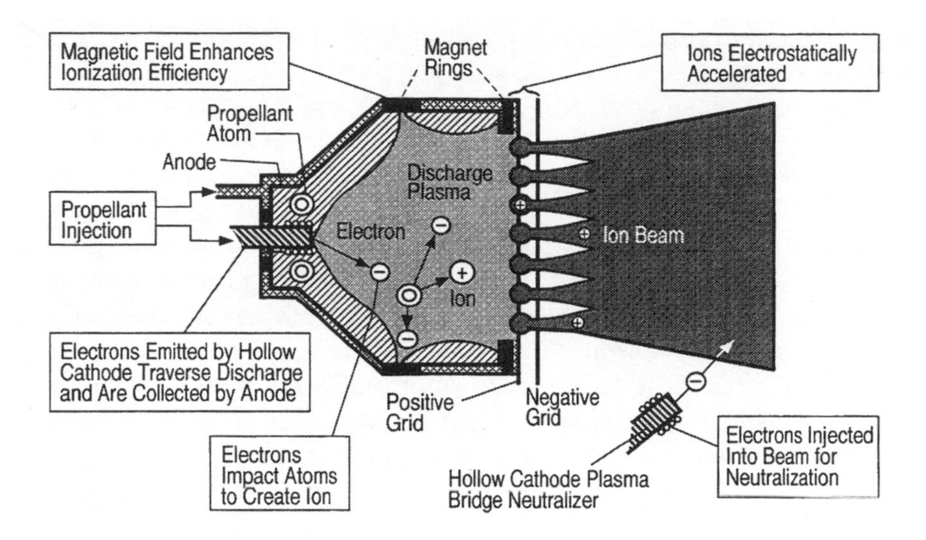

Figure 3.3. Schematic of an Ion drive. (Courtesy of NASA)

Figure 3.4. An ion thruster on the test stand. (Courtesy of NASA) (See also color insert.)

Earth surface gravities, so electric rockets must be deployed in space and fired for weeks or months to achieve high spacecraft velocities.

Unlike nuclear rockets and solar-thermal rockets, solar-electric rockets are now operational as prime propulsion for robotic interplanetary probes such as NASA's Deep-Space 1 and SMART-1, the European Space Agency's (ESA) first European mission to the Moon. SMART-1 was equipped with a type of electric propulsive device known as the Hall-effect engine, after a plasma phenomenon discovered by American physicist Edwin H. Hall in the 19th century. Solar-cell panels supplied power to the xenon ion engines, producing a thrust of about 68 mN, but operating for 7 months. The overall flight time to the Moon was about 14 months; during this time only 59 kilograms of propellant were consumed. The primary goal of this mission was not to reach the Moon, but rather to demonstrate that low-thrust, high-exhaust, velocity ion thrusters work very well in space as the primary propulsion source. ESA decided to extend the mission by more than 1 year, until September 3, 2006, in order to gather more scientific data.

Studies are under way in many countries that may soon increase the effective exhaust velocity of ion thrusters to 50 kilometers per second or higher.

So it may be surprising to the reader that electric rockets have so far been employed only for small robotic missions. Why have these reliable, high exhaust velocity engines not yet been applied to propel larger interplanetary ships carrying humans?

One problem is power. A lot more solar (or ultimately nuclear) power is required to ionize and accelerate the fuel required to accelerate a human-occupied craft massing about 100,000 kilograms than is required to accelerate a 200-kilogram robotic probe. But a more fundamental issue is fuel availability.

A number of factors influence ion-thruster fuel choice. First, you want a material that ionizes easily, so that most of the solar or nuclear energy can be used to accelerate fuel to high exhaust velocities rather than to sunder atomic bonds. Argon, cesium, mercury, and xenon are candidate fuel choices satisfying this constraint. But since space mission planners are also subject to environmental constraints, toxic fuels such as mercury and cesium are avoided in contemporary missions. Fuel storage during long interplanetary missions is also an issue—so contemporary electric rockets are fueled with xenon.

But if we propose an interplanetary economy based on large electric thrusters expelling xenon, we must overcome another issue. This noble (nonreactive) gas is very rare on Earth. Most of its commercial inventory is utilized for fluorescent lighting. Even a modest nonrobotic interplanetary venture would quickly exhaust the world supplies of this resource.

Nuclear-Direct: The Nonthermal Nuclear Rocket Concept

Although interstellar missions are not discussed until Chapter 9, in this section we briefly discuss a concept originated for interstellar flight in order to show some additional limitations related to rocket propulsion. In the 1970s, a number of investigators considered either nuclear fission or nuclear fusion for accelerating a spaceship to 0.1 c. The resultant one-way trip time between 40 and 50 years to Alpha Centauri was very appealing from the human lifetime viewpoint (35 to 40 years still represents a sort of minimum requirement for hoping to get approval for very advanced missions beyond the solar system). Here we comment on a concept (originated by author Vulpetti) that aimed at analyzing a multistage rocket starship exclusively powered by the nuclear fission.

Figure 3.5 may help us to figure out the central point of the nuclear-direct (ND) propulsion concept. Two types of fissionable elements are necessary in the form of two chemical compounds, say, FC1 and FC2 for simplicity. FC1 may be uranium dioxide or plutonium dioxide, whereas FC2 may be an appropriate compound of plutonium 239. They are stored in special tanks and supply two systems: a (so-called) fast nuclear reactor and a magnetic nozzle. The former one burns FC1 and produces an intense beam of fast neutrons, which are subsequently slowed down at the magnetic nozzle. Here, these neutrons induce fissions in FC1. The fission fragments and the electrons form high-energy plasma that is exhausted away through the magnetic field forming the nozzle. Why such a complicated arrangement?

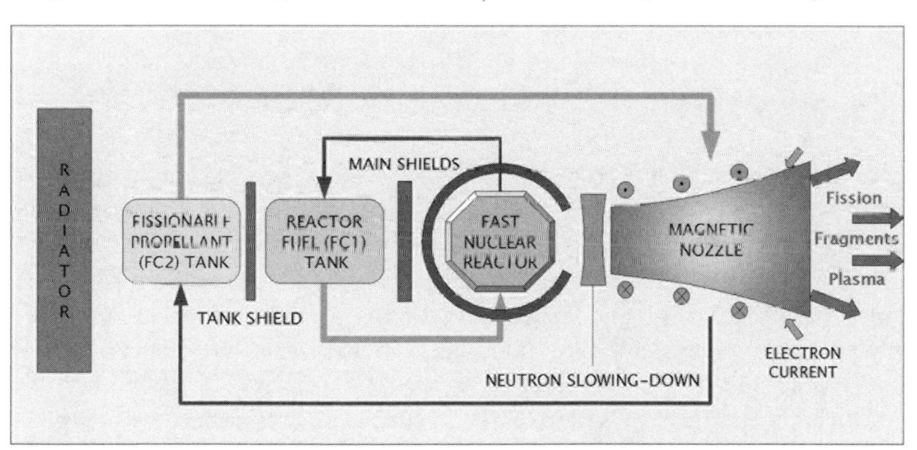

Figure 3.5. Conceptual scheme of a nuclear-fission engine exhausting the fission products directly, namely, using them as reaction mass. (Courtesy of G. Vulpetti)

The main reason is to utilize the enormous fission energy without passing through the production of heat to be transferred to some inert propellant like hydrogen. In other words, nuclear-direct would have avoided the exhaust speed limitations of the nuclear-thermal rocket (about 10–20 km/s). As a point of fact, the plasma from ND systems may be exhausted with a speed of 9000 to 10,000 km/s. Figure 3.5 presents an oversimplified schematic of the ND concept. Some of the related problems were analyzed quantitatively in the 1970s. Many major difficulties were found to relate to the practical realization of the reactor and the magnetic nozzle. The same concept has not been examined in the light of current knowledge about nuclear reactors, materials science, and sources of very strong magnetic fields. In any case, even if a multistaged starship of such a type were realizable by future technologies, the amount of fissionable elements to be managed would be so high that even the concept's author would be somewhat perplexed.

One should note that even a small-scale version of the ND concept would not be suitable for a human flight to Mars. Simply put, a crewed spaceship to Mars (and back) should have a rocket system capable of a jet speed of 20 to 40 km/s and an initial acceleration of 0.03 m/s^2. If one attempts to use a rocket with a jet speed 300 times higher, but using the same jet power per unit vehicle mass (in this case approximately 0.5 kW/kg), then the initial spaceship acceleration would be about 0.0001 m/s^2. Attempting to escape Earth—for a crew—with such an acceleration level would last months in practice and full of risk from radiations. So, one should go back to the nuclear-thermal rocket or the ion drive and solve the problems mentioned in the previous sections.

Nuclear-Pulse: The Ultimate in Rocket Design

Let's say that you're not content with slow accelerations and flights to Mars requiring 6 months or more, and let's also assume that the challenges of a nuclear-thermal single-stage-to-orbit do not go away. Instead, you become interested in the ultimate space voyages—across the 40-trillion kilometer gulf separating the Sun and its nearest stellar neighbors, the three stars in the Alpha Centauri system. Are there any rocket technologies capable of interstellar travel?

During the late 1950s and early 1960s, U.S. researchers pondered a remarkable, although environmentally very incorrect, rocket technology that was code named Project Orion (Fig. 3.6). In its earliest incarnations, Orion would have flown as either a single stage or a Saturn V upper stage.

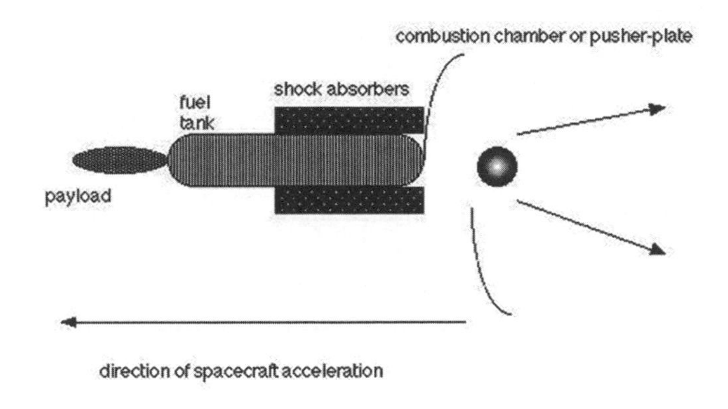

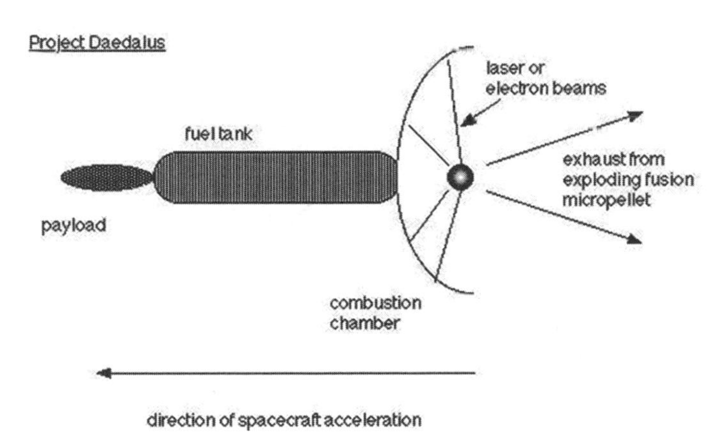

Figure 3.6. Two nuclear-pulse concepts. (From G. Matloff, *Deep-Space Probes*, 2nd ed., Springer-Praxis, Chichester, UK, 2005)

Orion passengers and payload would ride above the fuel tank, as far from the combustion chamber as possible. Fuel would consist of small nuclear-fission "charges" that would be ejected and ignited behind the main craft. Remarkably, materials exist that could survive the explosion of a nearby nuclear device.

Note the shock absorbers in Figure 3.6. These would ease the stress on the craft (and its occupants) from the uneven acceleration resulting from the reflection of nuclear debris.

On paper, Orion would have opened the solar system. Huge payloads could have been orbited by Saturn V with an Orion upper stage; this technology could have been used to perform rapid voyages throughout the solar system.

But Orion does not exist just on paper. Scale models, like the one on display in the Smithsonian Air and Space Museum in Washington, DC, flew through the air on the debris of chemical explosives and then parachuted safely to Earth.

As well as being a high-thrust device easily capable of launch from Earth's surface, the exhaust velocity of Orion's highly radioactive fission-product exhaust would have been 200 kilometers per second.

If the small nuclear charges were replaced with hydrogen bombs and if Earth-launched Orions were replaced with huge craft manufactured in space, perhaps using extraterrestrial resources, Orion derivatives could serve as true starships. In the unlikely event that the world's nuclear powers donated their arsenals to the cause, super-Orions propelled by hydrogen bombs could carry small human communities to the nearest stars on flights with durations measured in centuries.

But sociopolitical Utopia is a long way off. So, in the early 1970s, a band of researchers affiliated with the British Interplanetary Society began a nuclear-pulse starship study that was christened Project Daedalus. As shown in Fig. 3.6, a Daedalus craft would replace the nuclear or thermonuclear charges with very much smaller fusion micropellets that would be ignited by focused electron beams or lasers after release from the ship's fuel tank. The Daedalus fusion-pulse motor could theoretically propel robotic craft that could reach nearby stellar systems after a flight of a century or less. Larger human-occupied "arks" or "world ships" would require centuries to complete their stellar voyages. The proper propellant choice would greatly reduce neutron irradiation that would always be a problem for Orion craft. But major propellant issues soon developed.

The ideal Daedalus fusion fuel mixture was a combination of deuterium (a heavy isotope of hydrogen) and helium-3 (a light isotope of helium). Deuterium is quite abundant on Earth, but helium-3 is vanishingly rare. We might have to venture as far as the atmospheres of the giant planets to locate abundant reserves of this precious material.

The economies of Daedalus would be staggering. But they are nothing compared with the economic difficulties plaguing the ultimate rocket—one propelled by a combination of matter and antimatter.

A concept made popular by the televised science-fiction series *Star Trek*, the antimatter rocket is the most energetic reaction engine possible, with exhaust velocities approaching the speed of light. Every particle of ordinary matter has its charge-reversed antimatter twin. When the two are placed in proximity, they are attracted to each other by their opposite electric charges.

And when they meet, the result is astounding. In their interaction, all of their mass is converted into energy—far dwarfing the mass-to-energy conversion fraction of fission and fusion reactions (which never exceed 1 percent).

Antimatter storage is problematic. If even one microgram of antimatter fuel were to come in contact with a starship's normal-matter fuel tank, the whole complex would vanish instantly in a titanic explosion. Tiny amounts of antimatter, however, have been stored for periods of weeks or months, suspended within specially configured electromagnetic fields.

But what really dims the hopes of would-be antimatter rocketeers is the economics of manufacturing the stuff. A few large nuclear accelerators in Europe and the U.S. have been configured as antimatter factories. But an investment of billions of dollars and euros result in a yield of nanograms or picograms per year.

Someday, perhaps, solar-powered antimatter factories in space will produce sufficient quantities of this volatile material to propel large spacecraft at relativistic velocities. But until that far-future time arrives, we will have to search elsewhere to find propulsion methods for human-occupied starships.

Perhaps it's a good thing that cost-efficient antimatter manufacture is well beyond our capabilities. Imagine the havoc wrought by terrorists or rogue states if they had access to a nuclear explosive that could be stored in a magnetically configured thimble!

In ending this chapter on rocket's intrinsic limitations, we would like to make two points. The first one is conceptual. When one considers a very high nonchemical-energy density source (to be put onboard a space vehicle), there is always a basic difficulty in transferring energy from the source particles to the particles of the rocket working fluid. If one attempts to use the source's energetic particles *directly* as the exhaust beam, then one unavoidably has to deal with significant difficulties: the higher the particle energy, the more difficult it is to build a jet with a sufficiently high thrust.

The second point regards the context of spaceflight, in general, and space transportation systems, in particular. The design and function of small space engines, even though important for a spacecraft, are essentially of a technological nature. Quite different is the problem of a new space transportation technique, which also entails financial problems, safety and security issues, international cooperation (if any), long-term planning, and so on. Such problems are most obvious in developing a new launcher, which gives access to orbits close to the Earth. However, some difficulties arise even for in-space transportation systems to distant targets—not only for systematic human flights to other celestial bodies, but also for future scientific and utilitarian space missions, which will invariably increase in both complexity and number.

Further Reading

Many references describe the Apollo lunar expeditions of the late 1960s and early 1970s. For example, you may consult Eric Burgess, *Outpost on Apollo's Moon*, Columbia University Press, NY,1993. A more technical treatment is found in Martin J. L. Turner, *Rocket and Spacecraft Propulsion*, 2nd ed., Springer-Praxis, Chichester, UK, 2005. Turner's monograph also considers in greater detail many of the rocket varieties examined in this chapter.

Various nuclear approaches to interstellar travel are discussed in a number of sources. For a recent popular treatment, see Paul Gilster, *Centauri Dreams*, Copernicus, NY, 2004. A recent technical monograph is Gregory L. Matloff, *Deep-Space Probes*, 2nd ed., Springer-Praxis, Chichester, UK, 2005.

A photographic sequence showing an Orion prototype in flight is reproduced in Eugene Mallove and Gregory Matloff *The Starflight Handbook*, Wiley, NY, 1989. The history of Projects Orion and Daedalus are also reviewed in this semipopular source.

Non-Rocket In-Space Propulsion 4

Now that we've examined rocket theory, potential, and limitations, we are ready to consider some of the alternatives to this mode of propulsion. If our spacecraft is ground-launched, we might consider a jet as the first stage, where oxygen is ingested from the air instead of carried on board. Other launch alternatives include igniting the rocket while it is suspended from a high-altitude balloon or accelerating it upon a magnetically levitated track prior to ignition.

Although these alternatives are fascinating and well worth further study, we will not consider them further here. Instead, we will concentrate in this chapter on non-rocket alternatives that can alter the motion of a vehicle already in space.

Aeroassisted Reentry, Deceleration, and Orbit Change

Consider an Earth-orbiting spacecraft near the end of its mission that is ready to return home. We could simply fire the rocket in reverse and expend enough fuel to cancel the low-Earth-orbit velocity of 8 kilometers per second. At great expense in mission size and complexity, the craft would simply fall vertically toward Earth.

Very early in the space age, mission planners realized that such a procedure would be totally inadequate. Instead, they opted for aeroassisted reentries.

In an aeroassisted reentry, the spacecraft is first oriented so that a small rocket (a retrorocket) can be fired to oppose the spacecraft's orbital direction. The craft drops into a lower orbit where it encounters the outer fringes of Earth's atmosphere. Atmospheric friction further slows the craft so that it drops deeper into the atmosphere.

During an aeroassisted reentry, a spacecraft must be protected against the high temperatures produced by the frictional interaction between the vehicle

G. Vulpetti et al., *Solar Sails*, DOI: 10.1007/978-0-387-68500-7_4,

and atmospheric molecules. In many cases, an ablative heat shield is utilized. Ablation is akin to evaporation—small fragments of heat-shield material evaporate at high velocity, carrying away much of the frictional heat. Some spacecraft, such as NASA's space shuttle, use instead temperature-resistant ceramic tiles to protect the craft and crew during reentry.

When the craft has slowed sufficiently and descended further into the atmosphere, aerodynamic forces can be applied to control the craft's trajectory. Some returning space capsules—like Russia's Soyuz and China's Shenzhou—return to Earth ballistically with the aid of parachutes. Others, such as the space shuttle, are equipped with wings so they can glide to a landing. Some robotic craft—especially rovers bound for Mars—bounce across the surface on inflatable airbags after the descent.

In addition to the Earth, the planets Venus, Mars, Jupiter, Saturn, Uranus, and Neptune and Saturn's satellite Titan are equipped with atmospheres. Interplanetary robotic explorers have applied aerobraking at Mars and could apply this technique while orbiting other atmosphere-bearing worlds.

To perform an aerobrake maneuver, a spacecraft is initially in an elliptical orbit around an atmosphere-bearing world. If the low point of the orbit grazes that world's upper atmosphere and the spacecraft is equipped with a large, low-mass surface such as a panel of solar cells, it can utilize atmospheric friction on each orbital pass to gradually circularize the orbit, without the expenditure of on-board fuel.

A more radical maneuver is aerocapture (Figs. 4.1), which has not yet been tried in space. Here, a probe approaches the destination **planet** in an initial sun-centered orbit. Its trajectory must be very carefully calibrated and it must be equipped with a heat-resistant, low-mass, and large-area component that would ideally function like a parachute. In an aerocapture maneuver, one atmospheric pass is sufficient for the planet to capture the probe into a planet-centered orbit.

Planetary Gravity Assists: The First Extrasolar Propulsion Technique

Aeroassist is a fine non-rocket approach to deceleration. But how can a spacecraft increase its velocity without rockets? One approach, first used on the Pioneer 10/11 and Voyager 1/2 missions of the 1970s, is to transfer orbital energy from a planet to a spacecraft. Utilizing this technique, the Pioneers and Voyagers flew by the outer planets Jupiter, Saturn, Uranus, and Neptune, and have continued into the interstellar vastness beyond, as humanity's first emissaries to the Milky Way galaxy.

Figure 4.1. A rigid aeroshell could protect a payload during aerocapture. (Courtesy of NASA) (See also color insert.)

This maneuver works best when the spacecraft approaches a massive planet with a high solar-orbital velocity. Although Earth, the Moon, Venus (Fig. 4.2), Saturn, Uranus, and Neptune have been also been utilized, the best world in our solar system for gravity assists is Jupiter.

Let's say that you are planning a mission that will fly by Neptune and have a comparatively small booster. To maximize payload and not exceed your budget, you might initially consider flying a minimum-energy ellipse, with the perihelion at Earth's solar orbit (1 astronomical unit [AU]) and the aphelion at Neptune's solar orbit (about 30 AU). You'd better be patient and have a very young science team—the travel time will be about 31 years.

To save time, you will likely choose to inject the spacecraft into a Jupiter-bound minimum-energy ellipse, which requires a flight of only 2.74 years. You would then graze Jupiter appropriately to add velocity to the spacecraft and reduce its travel time to Neptune. This technique was utilized by Voyager 2, which required "only" 12 years to perform its Jupiter- and Saturn-aided flybys of Uranus and Neptune.

There are limits to gravity-assist maneuvers. If you approach a planet

Figure 4.2. The Venus gravity-assist performed by Saturn-bound Cassini in 1999. (Courtesy of NASA)

appropriately and your trajectory direction is altered by 90 degrees by the flyby, you can increase your spacecraft's Sun-centered velocity by about 13 kilometers per second—the orbital velocity of Jupiter about the Sun. If your craft arrives at Jupiter with very low velocity relative to that planet, your trajectory direction might be bent by 180 degrees. Then, you can increase spacecraft velocity by about 26 kilometers per second—twice Jupiter's solar-orbital velocity. In both cases, Jupiter will slow an infinitesimal amount in its endless journey around the Sun.

Another way to use the gravitational field of a large celestial body to accelerate a spacecraft is to perform a powered maneuver during closest approach to that body. Although such a powered gravity assist technically does not replace a rocket, it certainly increases the effectiveness of a rocket motor in altering a spacecraft's velocity. The most efficient **rocket**-powered gravity assist within our solar system would utilize a flyby of the Sun with the rocket ignited as close to the Sun as possible.

A general issue about planetary gravity-assist is that it depends strongly on the target planet's position; as a consequence, the launch window can be narrow, year-dependent, and low in mission repetition. Strictly speaking, gravity-assist is not a real propulsive mode: it is rather an advanced technique of celestial mechanics applied to spaceflight. It has been very fruitful in the past decades, but future astronautics needs devices that are also able to energize a space-vehicle almost continuously, far from any planet. Such systems do not exclude a mixed mode, namely, advanced spacecraft propulsion and gravity-assist combined.

Electrodynamic Tethers: Pushing Against the Earth's Magnetic Field

A tether is nothing more than a long, thin cable that attaches two spacecraft or spacecraft components (Fig. 4.3). If that cable is long (kilometers in length) and electrically conducting, then it can conduct electricity and use the interaction of that electricity with Earth's magnetic field to produce thrust. The physics is not complicated, but it is difficult to visualize. A current-carrying wire generates a magnetic field. Conversely, a wire moving through a magnetic field produces a voltage difference across the length of the wire. If electrical charge is available at one end and the circuit "closes" back with the other end, then a current will flow across this potential difference and through the wire. If the wire happens to be moving through space around Earth, then it is moving through a magnetic field. (For proof, just get out your compass to see the effects of Earth's magnetic field.) In low Earth orbit, there are lots of ions and electrons to provide the current (in what is commonly called the *ionosphere*) and the circuit closes with only one wire being used by virtue of electrical conduction through the ionospheric plasma created by these same ions and electrons. The current flowing through the wire tether then experiences a force due to its motion through the magnetic field. This force is perpendicular to both the **local** magnetic field and to the direction of current flow. Since the current is trapped inside the wire, the force effectively pushes on the wire in either its direction of motion through space (accelerating it) or, if the current is flowing in the opposite direction, decelerating it.

The tethered satellite system and plasma motor generator missions of NASA demonstrated the electrodynamic properties of tethers in space in the

Figure 4.3. A satellite moves toward a higher orbit after release from a tether-equipped spacecraft. (Courtesy of NASA) (See also color insert.)

1990s. The use of electrodynamic tethers for propulsion in space remains to be demonstrated.

Momentum Exchange Tethers: King David's Slingshot To Space

Using a tether to exchange momentum or orbital energy between two spacecraft was demonstrated in space by the flight of the small expendable deployer system (SEDS) missions in the mid-1990s. The SEDS-1 mission saw the deployment of a 20-kilometer nonconducting tether from the upper stage of a Delta-II rocket after its primary mission was complete. The SEDS deployed a tethered, spring-ejected, 25-kilogram end mass (basically a deadweight) from the Delta-II stage. The spring gave the end mass enough of a "kick" to move to a distance where the *gravity gradient* **(see Glossary)** took over, resulting in the end mass being fully deployed 20 kilometers from the stage. After reaching its full 20-km length, the tether was cut at the deployer, sending the end mass to reentry. This technique might be used to boost the orbit of valuable space assets, while assisting others to reentry—all without the expenditure of propellant.

A momentum-exchange tether might work like this. Let's say that you are expedition commander aboard the International Space Station (ISS). Periodically, you must schedule a short rocket burn to counter atmospheric drag and maintain the station's orbit. If you happen to have a long momentum-exchange tether, say one that is 50 to 100 kilometers in length and a space shuttle has checked in for a visit, you can cancel the thruster burn and save fuel. All you must do is to attach the shuttle and the ISS with the tether, position the shuttle below the tether, and slowly unravel the tether. When the shuttle is at a sufficient distance below the ISS, sever the connection. The shuttle will drop to a lower orbit and position itself for reentry; the ISS will soar to a higher orbit.

The main issue delaying operational application of these devices is safety. Mission directors are concerned about the orbital debris that could result from a malfunctioning tether.

But tethers could be applied all over the solar system, wherever there are gravitational and magnetic fields and sources of electrons. This technology may even have interstellar applications.

MagSails and Plasma Sails: Riding the Solar Wind

In addition to electrodynamic tethers, there are other propulsion **concepts** that use purely electromagnetic interactions instead of rocket-based momentum exchange to derive thrust. Two of these are the MagSail and its cousin, the plasma sail.

A MagSail uses the magnetic field generated by a large superconducting wire loop to reflect solar wind ions. These ions, generated in copious amounts by the sun, stream outward into the solar system. Using the same principles as an electrodynamic tether, that a magnetic field exerts a force on a charged particle or collection of them (as in an electrical current), the MagSail would use the magnetic field generated by the current flowing through the superconducting wire coils to reflect solar wind ions back in the direction from which they came, transferring their initial energy and momentum to the MagSail.

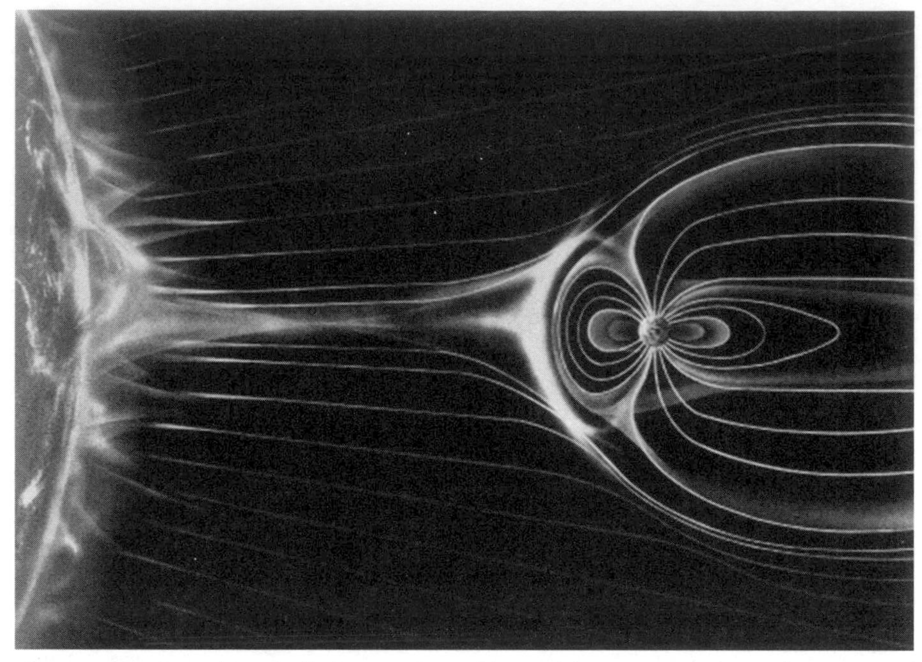

Figure 4.4. The solar wind's interaction with Earth's magnetosphere. (Courtesy of NASA) (See also color insert.)

In reality, things are very complicated. Not only is the reflection of solar protons not guaranteed at any distance from the sun, but also the flux of the incoming protons changes considerably during days, months, and years. Since we cannot control such fluctuations, a big problem arises because of the difficulty in ultimately controlling the spacecraft motion.

Dr. Robert Winglee at the University of Washington proposed another version of the MagSail. The Winglee concept is called mini-magnetospheric plasma propulsion (M2P2). The M2P2 would function in space by creating a small-scale version of Earth's magnetosphere. (The magnetosphere is defined as the region of space near Earth in which electromagnetic interactions are controlled by Earth's magnetic field. It includes the ionosphere, Fig. 4.4) The M2P2 would generate an artificial magnetosphere in which trapped electrically charged particles would reflect the solar wind over a wide area (many kilometers). This would, in theory, allow an attached spacecraft to accelerate outward from the Sun without the expenditure of rocket fuel. Early experiment and analysis are inconclusive regarding the overall feasibility of the technology, but stay tuned. The idea is still in its infancy. Even if M2P2 fails as a non–rocket-propulsion device, it may be useful in protecting astronauts in deep space from cosmic rays by acting as a large deflecting screen.

Interstellar Ramjets and Their Derivatives

Speaking of interstellar applications, this chapter would be incomplete if we did not mention the most dramatic rocket alternative of them all. This is the interstellar ramjet (Fig. 4.5). When first proposed by the American physicist Robert Bussard in 1960, it seemed to demonstrate an economically acceptable method of achieving spacecraft velocities arbitrarily close to the speed of light.

Even though interstellar space is a very perfect vacuum by terrestrial standards, it is far from completely empty. As well as the occasional dust particle, galactic space is filled with a diffuse (mostly) hydrogen gas with an average density of about one particle per cubic centimeter. Bussard proposed a spacecraft that would fly through this medium at high speeds. Utilizing electromagnetic fields, it would ingest interstellar hydrogen, probably in the ionized form of protons and electrons, and funnel this material into a thermonuclear fusion reactor very far in advance of any technology we can dream of. Instead of the comparatively easy reactions between heavy hydrogen isotopes and low-mass helium isotopes that fusion researchers experiment with today, this reactor would burn hydrogen directly to obtain helium plus energy, thereby duplicating the energy-producing process of the Sun and most other stars.

Because there is no on-board fuel, the ramjet's ideal performance is limited only by its mass and the density of the local interstellar medium. Under optimum conditions, it could accelerate constantly at one Earth gravity. The interstellar ramjet would approach the speed of light after only

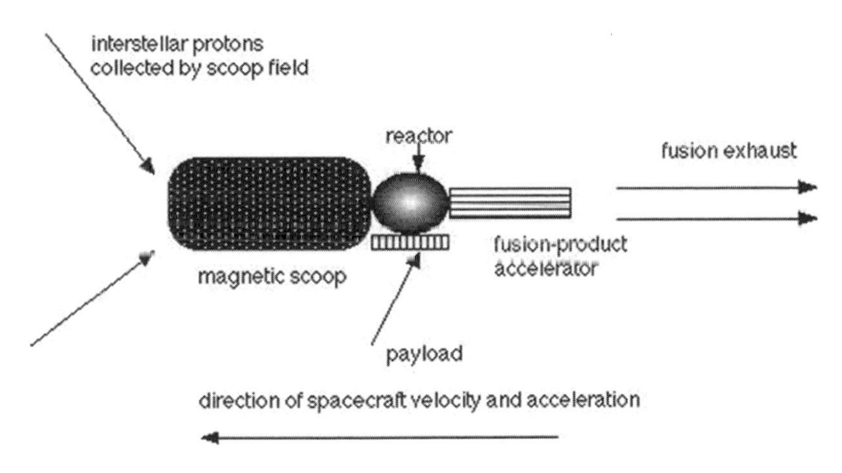

Figure 4.5. Bussard's proton-fusing interstellar ramjet concept. (From Gregory L. Matloff, *Deep-Space Probes*, 2nd ed., Springer-Praxis, Chichester, UK, 2005)

about one year of operation. Due to relativistic time dilation, the on-board crew could reach very distant interstellar destinations within years or decades from their point of view, although much longer time intervals would pass from the viewpoint of stay-at-home Earthlings.

Almost immediately, the interstellar ramjet became the darling of science fiction. If a star-bound astronaut invested her salary before launch and collected compound interest, would she own Earth upon her return a century later? If a malfunction occurred, might a time-dilated star crew keep accelerating and witness the final heat death of the universe and a new Big Bang, as happens in Poul Anderson's classic tale *Tau Zero*?

Alas, issues for the ramjet soon emerged, which dimmed the initial enthusiasm. From an astronomical viewpoint, it became apparent that our solar system resides in a vast galactic bubble of interstellar gas with a density of less than 1 hydrogen atom or ion per 10 cubic centimeters, far smaller than the average interstellar gas density. This implied that an enormous electromagnetic scoop radius would be required—in the vicinity of thousands of kilometers—for even a reasonably massed starcraft.

But physics was no friendlier to Bussard's wonderful dream. The required proton–proton reaction is not only a bit more difficult to ignite than currently feasible fusion reactions, but it is many orders of magnitude more difficult! Indeed, to fly to the stars using an interstellar ramjet, we might need a star to ignite the interstellar hydrogen—not the most mass effective of interstellar propulsion modes!

So a number of less challenging derivatives of the ramjet concept have been introduced. One is the ram-augmented interstellar rocket (RAIR). This theoretical "ducted" rocket carries fusion fuel as its energy source. It could, in principle, increase the efficiency of its fusion-pulse rocket by adding collected interstellar ions to the exhaust stream. If we replace the on-board fusion reactor with a receiver of laser energy beamed from the inner solar system, on-board fuel requirements are greatly reduced.

Another **concept** is the ramjet runway, in which a trail of fusion-fuel pellets is deposited in advance of the accelerating starship, which collects, reacts, and exhausts the pre-deposited fuel. Another possibility is to utilize the electromagnetic scoop to slowly and gradually collect **fusible nuclei** from the solar wind prior to an interstellar fusion rocket's departure.

All of these approaches have their own developmental issues. Although they are not as efficient nor elegant as the pure ramjet, at least they offer some hope to designers of future interstellar spacecraft.

But even if the physics problems in the construction of certain ramjet derivatives may not be insurmountable, there are major technological issues. Setting aside the major issues involved in fusion-reactor design, it must be noted that construction of electromagnetic interstellar ion scoops is far from straightforward.

Early scoop concepts were generally developed analytically during the 1970s. Further analysis with plasma-physics computer codes revealed that most scoop concepts would tend to reflect interstellar ions rather than collect them. In other words, an electromagnetic scoop field would serve better as an interstellar drag brake rather than as an aid to non-rocket acceleration.

Most of the above propulsion concepts are particular cases of the Multiple Propulsion Mode (MPM), a concept introduced by author Vulpetti in 1978 and improved in 1990. Such a mode does not entail a multistage space-vehicle necessarily. Its principle is different: if one uses rocket, ramjet, and laser-sail in a special configuration of *simultaneous* working and sharing the total power available to a (huge) starship, then truly relativistic speeds could be achieved. As a point of fact, it has been proved mathematically that these three propulsion systems may be made equivalent to a single rocket endowed with an exhaust beam of almost the speed of light, but with a thrust enormously higher than that obtainable for a relativistic photon rocket. The mass of such a starship may be greatly lessened only if antimatter were used as the rocket fuel. However, the antimatter amounts for reaching nearby stars would be much, much higher than our current production cabability. Nevertheless, MPM studies show implicitly that even utilizing advanced concepts of current physics, *fast* interstellar travel is completely out of our current or medium-term capabilities; although the MPM is conceptually clear, a real MPM-based starship would be so complex that, simply put, we do not know how to engineer it. Different approaches are necessary for the interstellar flight, including an appropriate enlargement and understanding of our present physics.

Further Reading

The kinematics of minimum-energy or Hohmann interplanetary transfer ellipses are presented in many technical books on astrodynamics. One good source is Roger R. Bates, Donald D. Mueller, and Jerry E. White, *Fundamentals of Astrodynamics*, Dover, NY, 1971.

For a technical review of planetary gravity-assist technology, interstellar ramjets, MagSails, and the mini-magnetosphere, consult Gregory L. Matloff *Deep-Space Probes*, 2nd ed., Springer-Praxis, Chichester, UK, 2005. A more popular review of these topics is Paul Gilster, *Centauri Dreams*, Copernicus, NY, 2004.

G.L. Matloff, L. Johnson, C. Bangs, *Living Off the Land in Space*, Praxis-Copernicus, 2007. Aeroassist, tethers and related technology are treated in this book as well.

For the multiple propulsion mode concepts, papers were published in the *Journal of the British Interplanetary Society* (JBIS) in the 1970s and 1990s.

The Solar-Sail Option: From the Oceans to Space

In the previous chapters, we described the space rocket engines, how they work, their role in past and current spaceflight, and their limitations. We have also shown that the rocket is not the only propulsion type that could be employed in space. Among the types of space propulsion currently under investigation, one is particularly promising: the solar sail. This propulsion mode is not conceptually new, even though only recent technology gives it a good chance to make a quality jump in spaceflight. Its principles and how to efficiently use a sail vehicle could be understood better by reviewing what happened about four millennia ago on the seas and by referring to the progress of physics in the 19th and 20th centuries. Early pioneers of solar sailing conceived a space use of sails in the first half of the 20th century, whereas the first technical publications and space designs began in the second half. But let us proceed in chronological order.

A Bit of Human History

Well before physics was founded as a branch of scientific knowledge, human beings needed to travel on water: rivers, seas, and oceans. Modern studies about human history and its evolution have shown that many peoples migrated through oceans, too, although the ancient boats were built and applied empirically. In particular, about 25 centuries B.C. a Middle East zone, corresponding to modern Lebanon, Israel and part of Syria, saw the first settlements of a famous people, mentioned in the Bible and named the Phoenicians after Homer.

Although Phoenicia consisted of many city-states (Sidon, Tyre, Beritus [now Beirut], Tripoli [near Beirut], Byblos, Arvad, Caesarea, etc.), they considered themselves one nation, which had a significant impact on their evolution. At their beginning, Phoenicians navigated using rafts. Then, they built more sophisticated boats that were used for fishing and coasting, but only in the daytime. Phoenicia occupied a geographic position strategic for

G. Vulpetti et al., *Solar Sails*, DOI: 10.1007/978-0-387-68500-7_5,

land and sea commerce in the context of the Egyptian and Hittite empires. However, Phoenicians should not be confused with Anatolian Hittites, originating from Anatolia (peninsula of Turkey) 19 centuries B.C. This Hittite empire was at its maximum expansion in 14th century B.C. (Anatolia, northwestern Syria, and northern Mesopotamia), while Phoenicians were under the control of Egypt. As a point of fact, Egypt conquered Phoenicia about 18 centuries B.C. and controlled its city-states until about the 11th century B.C. (Some scholars suppose that even the biblical Hittites were different from Phoenicians.)

Once independence was restored, Phoenicians began their expansion, not excessively in terms of military conquest, but above all as the most skilled navigators and clever merchants of the whole ancient world. In practice, they invented the modern concept of commerce and exploited its power. Their fleets were employed to expand commerce, explore new lands, found new colonies, and to transport precious goods.

According to the ancient Greek writers, Phoenician sea-craft resulted in two kinds of somewhat sophisticated large boats: war vessels and merchant ships. With their ships, Phoenicians not only dominated the Mediterranean Sea, but also traversed the Pillars of Hercules (now called the Straits of Gibraltar) and entered the Atlantic Ocean, reaching as far as the Canary and Azores Islands; they traveled past the coast of France and reached Wales, where they controlled the tin market.

Herodotus wrote that Phoenicians managed to circumnavigate Africa and pioneer a way to Asia about six centuries B.C., but there is no further witness regarding this extraordinary adventure. Around 950 B.C., King Solomon entrusted Phoenician crews for a commercial mission onto the Red Sea, whereupon their ships are said to have reached the southwest parts of India (actually, some relics of their journey has been found there). In the ninth century B.C., Phoenicians founded Carthage, located close to modern Tunis. That powerful city, which continued the commercial activities after Phoenicia's decline, was destroyed by Rome after three wars lasting more than a hundred years.

How, in those ancient times, was such extended exploration and trade possible? Let us list the key points. In the course of time, Phoenician ship builders, whose knowledge was based almost exclusively on experience rather than analysis, endowed their ships with a well-designed keel, different from the ships of other nations. Ships were long, streamlined, and exhibited two or three lines of oars (with related rowers) on different planes (biremes, triremes). In addition, a Phoenician ship had a mast with one sail. Although sails were of the square type and inappropriate for moving upwind (e.g., close hauled, close reach, beam reach), Phoenicians utilized them

Figure 5.1. Outline of Phoenician merchant vessel.

intelligently. At one point in their nautical history, they replaced one line of merchant-ship rowers by a sail, thus achieving three objectives:

1. The weight and volume of rowers and oars were occupied by a significantly heavier payload,
2. Travel duration decreased,
3. The vessel was endowed with two complementary propulsion systems—rowing and sailing—which were engaged according to the winds or on contingency. Another key point was that they learned to navigate by stars, especially the Polar Star, and to plot a course!

Figure 5.1 shows an outline of a Phoenician merchant vessel. Sailing, as developed by Phoenicians, was the foundation for all the improvements and new discoveries that allowed big vessels to navigate extensively on the seas and the oceans of the world. Today, the sail is still used as the main propulsive engine in a few places. Though replaced by steam, oil, and nuclear engines, water and land sailing are very much alive as sports. Looking back on previous millennia, one can assert that sailing has not been the ultimate maritime propulsion, but has represented a period of transition until knowledge made another quality jump possible. Many things were accomplished in such a transition! We have more than one reason to believe that similar things could and should happen in space.

Again, to better grasp space sails and what one could do with them, it is appropriate to devote a few words to sea sailing. In particular, we emphasize the roles of the different forces acting on the sail and the boat. It will be useful to find analogies and differences when we discuss the many types and applications of space sailcraft.

Sea Sailing

Even a small boat with one sail is a very complicated system. Here, we are concerned with the forces that allow a sailboat to move and remain stable. We shall use an oversimplified picture for illustrating them, but maintain qualitative correctness.

Just A Few Words About Wind

We sense wind as a mass of air moving from some direction. For most purposes, such movement can be considered almost horizontal. Winds are classified by either the kinds of forces acting on the air, their movement scale, or the geographic zones where they periodically occur or persist. There are a great variety of winds.

The main physical causes of air motion are the variable solar heating at different places and Earth's rotation. The former causes a difference of pressure between the air parcels at different places. Also important is the distance between the points with different pressure. A nonzero pressure difference is the main driver for moving a fluid (a gas or a liquid). The abundant energy that the Sun delivers to Earth causes many important phenomena, including the regional temperatures and pressures we hear about from television or read on Internet weather forecasts. Winds are powered by transformed solar energy, which is absorbed by atmosphere, lands, oceans, and on forth. If there were no solar radiation impinging on Earth, there would be no wind. As we shall see in the next section, a terrestrial sailboat utilizes this clean solar energy. We shall one day do the same thing with space sails.

How Can a Sailboat Navigate?

Although many small sailboats resemble the one in Figure 5.2, with two sails that behave as one combined sail, to understand sailing, it is sufficient to discuss one boat with one sail. As well as the sail, one important thing to be noted is the centerboard, which serves as the keel on larger sailboats.

Figure 5.2. A two-sail boat.

The velocity of a body is a vector, namely, a mathematical object with magnitude and direction of motion. (Accelerations and forces can be pictured by vectors as well.) Often, in physics, the magnitude of the vector velocity is called speed. If an observer sees two bodies with the same (vector) velocity, this means that both bodies have the same speed and direction of motion with respect to the observer. (The observer can be a scientific device as well.) There exists no absolute velocity. Some frame of reference has to be defined. However, sometimes when the phenomenon (to be studied) includes many different bodies, one may call the velocity of the body B_j (i.e., the body labeled with "j") with respect to observer O (who is at rest with respect to some main reference frame) the absolute velocity of B_j, or $\mathbf{V}_j$. If another observer is connected solidly to B_k, then the observer sees B_j, moving with respect to B_k, with the relative velocity $\mathbf{V}_j - \mathbf{V}_k$. If you remember your high school science, you may recall that such difference follows the rule of parallelograms for summing two vectors. In fact, it may be written $\mathbf{V}_j + (-\mathbf{V}_k)$, where the term in parentheses means "the opposite of $\mathbf{V}_k$."

Let us describe a simple but meaningful case: a boat with no sail, or other propulsive device, is floating on a water stream moving with respect to the land with a speed of 10 knots (1 knot equals about 0.51 m/s) and parallel to the shore, for instance from left to right as observed by a person on shore and facing the sea. We can denote such boat by B1, the sea stream by B2, and the person by P1. In addition, this observer senses a wind (B3) with the same velocity. Let us assume that a sailor (S1) is simply seated on B1. Apart from seeing the land moving from *his* left to right, what else does he observe? S1 senses both water and air at rest. Furthermore, the *relative* velocity between water stream and wind (B2 and B3) is zero as well (according to the technical note in the above box).

> The direction of a plane (or an almost-plane) surface is that of the line perpendicular to the surface itself. For instance, by saying that a sail is downwind or upwind means that it is perpendicular to the wind velocity.

Now, suppose that the sailor raises a small sail. What happens to the boat motion? Nothing, of course, since no wind, as seen from S1 onboard, is blowing on the sail. Instead, let us suppose that P1, on the land, observes an increase of the wind's velocity up to 15 knots. Let us move to sailor's viewpoint. S1 senses an *aft* wind of 5 knots parallel to the shore, the wind fills the sail, pushes the boat, and increases the boat speed with respect to the sea, namely, it now creates a nonzero *relative* velocity between the boat and the water. (Implicitly, here we have supposed that the sail is downwind-oriented).

We are not finished. In addition to the action (or force) of the wind on the sail, there is another action (or force) on the hull, because B2 is now striking that part of B1 immersed in the water. In this simple example, the action from the water consists only of an increasing resistance, called the drag, to the sailboat motion. As the boat speed increases, the drag balances the force on the sail progressively: when the vector sum of these contrasting forces vanishes, the sailboat cruises. All this takes place because, as pointed out above, water and air have a nonzero relative velocity, which ultimately allows the boat to *control* its motion. Everybody who wants to go offshore on a small sailboat knows (at least intuitively) that no sailing is possible if air and water do not move with respect to one another.

What happens if the sail is at an arbitrary angle with respect to the direction of the wind sensed onboard? In general, we can have a complicated situation, but let's keep it simple. If the sail is oriented like that of Figure 5.3, then it behaves similarly to a well-shaped vertical wing. The pressure over the sail's downwind surface is higher than the pressure over the sail's upwind surface. Thus, a net force (called the *lift*) acts on the sail; it actually is a lateral

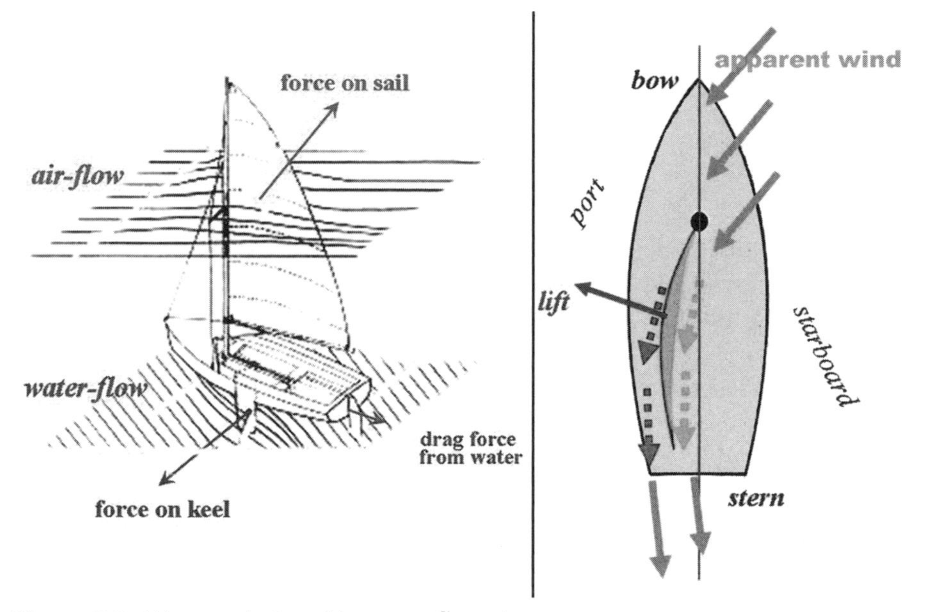

Figure 5.3. Water, wind and boat configuration.

or leeway force with respect to the boat stern–bow axis. A problem arises. Although there is a large component of the force along the longitudinal axis of the sailboat, there is also a significant leeward component. What prevents the sailboat from experiencing an uncontrolled sideways drift? Here is the role of the water. If the boat is endowed with a keel (or its small version, the centerboard), namely, a large quasi-flat surface under the boat, then the strong resistance of the water on this surface prevents the whole boat from being moved transversally. Thus, sailors can sail. Figure 5.3, on the left, shows the main forces: the (aerodynamic) force on the sail, and the (hydrodynamic) forces on the keel and the hull. The sailboat can accelerate along the course determined by the sum of such forces until this total vanishes. Afterward the sailboat can cruise. If the skipper turns the rudder or the wind changes, then new cruising-equilibrium conditions arise.

Figure 5.4 shows that the various forces are applied to different points. For instance, the lift is applied to the sail pressure center, say, P, whereas the force on the keel can be envisaged as "concentrated" on its pressure center, say, K. Let us look at Figure 5.4 showing the boat from the stern/bow. Since P and K do not coincide, there is a torque that causes the sailboat to tilt leeway. This is the *heeling torque* that would cause the sailboat to capsize, if unbalanced.

Is there anything that may halt this otherwise progressive tilting? Yes, there is. As a matter of fact, two other significant forces act on the sailboat. One is its weight, downward-applied to the center of mass (or the

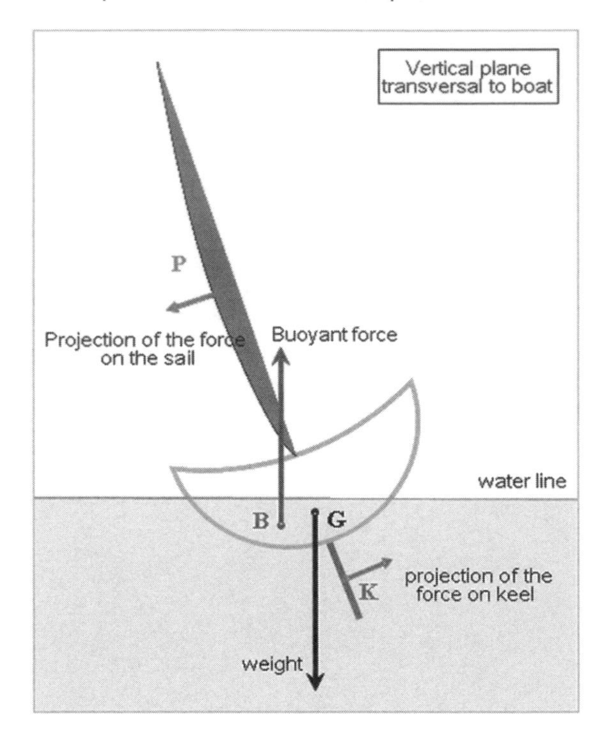

Figure 5.4. Main forces on a sailboat. (Courtesy of G. Vulpetti)

barycenter), say, G of the whole boat, including any crew. The other one is the Archimedes force, or the buoyancy, which is upward-applied to the barycenter, say, B of the water mass *displaced* by the hull. B is also called the center of buoyancy. The pair of these forces generates a torque that rotates the sailboat in the opposite way from the heeling torque, namely, the weight-buoyancy pair forms a countertorque. Therefore, besides the balance of the forces themselves, one has to maneuver to also achieve the balance of the above torques. As a result, the sailboat can cruise stably upwind (but not *directly* upwind) in a configuration similar to that shown in Figure 5.4.

The orientations of both sail and sailboat are complicated, but quite necessary to navigate and change course. Although the forces acting on a space solar sail are of different kinds, the orientation of a space sail shall be controlled accurately to obtain the desired path toward any target far from Earth.

Early History of Space Sailing

If rockets are inappropriate (and they are indeed) for advanced and long-range missions in the future spaceflight, one is required to search for valid propulsion alternatives. As previously discussed in Chapter 4, there are many potential non-rocket propulsive devices. That means that external-to-vehicle sources of energy are used; otherwise the spacecraft would be a rocket (as explained in Chapter 2).

Where would one find an external powerful source with plenty of energy free of charge? It is natural to resort to the nearest star, which is not Proxima Centauri! It's the Sun. Our Sun is really a stupendous engine producing energy. Deep inside the Sun, an enormous amount of hydrogen (over four million metric tons every second) is continuously transformed into helium via nuclear fusion. The total energy stored in the solar core is so high that, if the fusion were halted now, the Sun would remain luminous for at least twenty million years. On the other hand, the energy produced by the solar nuclear reactions in a given time interval requires thousands of years to reach the visible solar surface (also known as the photosphere).

Above the photosphere, there are a thin layer named the chromosphere *and* above it an extended zone, known as the solar corona (much lower in density than the photosphere, but much hotter). Above the photosphere, sunlight is essentially free to move through space in every direction. The corona produces additional light, which propagates into space. As a result, our star emits abundant ultraviolet light; visible, infrared, and—in much lower amounts—radio waves; and x-rays. We shall see in the next section that all forms of light are of the same nature—**the photon**—solely differing in energy. All photons are massless. Here, we call this overall "stream" of light the *solar photon flow.*

In addition, the Sun releases plenty of massive particles, essentially protons and electrons (95 percent), alpha particles (4 percent), and other ionized atoms. Such particles form what is known as the *solar wind* and should not be confused with the solar photon flow in any case. For spaceflight purposes, say that solar wind is inhomogeneous, variable, and largely unpredictable. When the solar wind approaches a body endowed with a magnetic field (like Earth and some other planets), the interaction is most complex and dependent in part on the Sun–body distance. Thus, it seems improbable that controlled spacecraft trajectories may be obtained, for months or years, by something so variable. Here is the first important difference with sea-sailing: we shall not use a wind of matter particles, such as the usual wind, for pushing space sails.

Then, may we utilize the solar photon flow for space sailing? If so, what should we do in practice? The latter question is largely related to various technological developments. In contrast, the former question is more conceptual and is related directly to the physics of light.

The Amazing Nature of Light

Light was the object of very active investigation by some ancient natural philosophers, long before the physicists of the modern era. Here are some examples. In the first century B.C., Lucretius asserted that the solar light and heat consist of corpuscles; he pursued the basic idea of Democritus (400 B.C.), and his followers that anything is composed of indivisible minute particles that he called *atoms*. Ptolemy dealt with the phenomena of reflection and refraction of light. His assumptions were expanded and improved upon by the Persian mathematician Alhazen, who wrote many books on optics and tried to verify his theory by experiments around 1000 A.D.! His major work was translated into Latin in 1270, but published in Europe only in 1572, twenty years before the birth of Pierre Gassendi, the great scientist and philosopher who also insisted on the atomistic view of light.

Isaac Newton started from Gassendi's work and oriented himself with the corpuscular view of light. Newton's theory on light appeared somewhat consistent with the luminous phenomena known at that time (i.e., reflection and refraction). Newton knew that his theory was unable to explain *partial* reflection, but he did not worry excessively. Newton's view about the nature of light dominated physics for many decades in the 18th century. Nevertheless, in the second half of the 17th century, Robert Hooke and Christian Huygens published their work (30 years apart) about a wave theory of light. This theory was able to predict the significant phenomena called the *interference* and the *polarization* of light. Thomas Young showed experimentally, by focus on the *diffraction* of light, that light behaves as waves. Euler, Fresnel, and Poisson argued that the phenomena related to light could be more easily explained by special waves propagating in a medium filling all space (the ether), obeying the principles contained in the theory. Even partial reflection could be explained by the wave theory of light, *provided that* the light involved is sufficiently intense.

In 1873, James Clerk Maxwell published *A Treatise on Electricity and Magnetism*. By a stupendous conceptual synthesis of the previous ideas and experimental results, including those connecting light to electromagnetism (Michael Faraday, Armand Fizeau), he derived the fundamental equations that still bear his name governing the behavior of the electric and magnetic forces in space, in time, and inside matter. In electromagnetic theory, light consists of high-frequency waves that are able to propagate in vacuum and in any direction at an ultra-high (constant) speed (see *speed of light* in the

Partial reflection is not an unusual phenomenon observed only in the laboratory, but is very common indeed. If you have a transparent medium (water, glass, Plexiglas, and so on) and a source of light (directed out of it) on the same side as you, you can be sure that most of the emitted light passes through the medium. However, a small fraction of it is reflected from the surface nearest you, hence its name. (Examples of this phenomenon include the moonlight reflected from a lake, and looking at shop's window allows one to see objects behind the glass as well as the reflected images of external objects.) In addition, you can note the same thing from a second subsequent surface, and so forth, of the medium or set of media. If you arrange a laboratory experiment where you can use many slabs (usually, one at a time) of different thickness, a source of one-color light (e.g., a red laser like the pointer of a lecturer), and a detector of light, the experimental outcomes are really strange, so strange that the phenomenon persists even when the source gets dimmer and dimmer. Only quantum physics can explain this phenomenon fully.

Glossary). The electromagnetic theory asserts that radio waves, thermal radiation, and visible light all are of the same nature: the only difference is in the vibration frequency or, equivalently, the wavelength. (How the electromagnetic range or spectrum can be subdivided will be discussed in Chapter 15.)

Scientists of the 19th century accepted Maxwell's theory: every thing about light was apparently and elegantly explained. Plenty of experiments were arranged in the subsequent years for testing electromagnetic theory. Thus, the particle view of light fell into disrepute.

However, some problems were appearing on the horizon. Let us mention one having basic relevance to sailing in space. In his treatise, Maxwell pointed out that if waves propagate in a certain medium, then one should expect a pressure on a surface perpendicular to the propagation direction. In 1873–74, William Crookes claimed to have found the experimental proof of the radiation pressure by means of his radiometer. However, a careful analysis by many scientists showed that he was measuring a different phenomenon. In 1876, independently of Maxwell, Italian physicist Adolfo Bartoli showed, by a *conceptual experiment* (see Glossary), that the celebrated second law of thermodynamics, which the whole Universe obeys, requires that the light emitted by a body, at any temperature, must exert a pressure on a material surface.

Although very small, as predicted by the Maxwell theory, the pressure of light represented a crucial point because of its profound meaning. In 1887, Heinrich Hertz proved the existence of electromagnetic waves experimentally; he showed that such waves can travel through space and set some theoretical foundations that Guglielmo Marconi subsequently extended and applied to his invention of the radio. Finally, in 1900, Russian physicist Pyotr Lebedev measured the pressure of light on a solid surface by means of an ingenious device. That was possible also because of the advancement of such technologies as intense light sources, pumps capable of producing near-vacuum conditions in a small volume, and so on. Lebedev's work was confirmed by U.S. physicists Ernst Fox Nichols and Gordon Ferrie Hull in 1901 by means of a radiometer. Thus, this prediction of electromagnetic theory was confirmed.

After Max Planck's hypothesis was published in 1900, in 1905 (the same year as special relativity), Albert Einstein explained all features of the photoelectric effect by quanta of light (see box, below). Thus, light appeared to be an electromagnetic wave endowed with the particle properties of energy and *momentum* (see Glossary) in *discrete* amounts! In 1926, U.S. chemist Gilbert Lewis proposed the term *photon* for the electromagnetic quantum of light.

In 1900, Max Planck published the first quantum theory, in which light is treated as a particle endowed with *discrete* amounts of energy, called quanta. Planck considered such particles to be the basic units of energy. According to Albert Einstein, the energy of such an element is expressed by $E = h\nu = hc/\lambda$, where λ denotes the wavelength. Although very simple, this relationship contains two of the most important constants of nature, the speed of light (c) in vacuum and the *Planck constant* (h) (see Glossary). In addition to energy, such elements transport *momentum* (p) (see Glossary) given by $p = E/c = h/\lambda$. Even though it is the carrier of energy and momentum, the quantum of light does not have mass!

In the course of the 19th century, many scientists worked experimentally on another strange phenomenon: when ultraviolet light or x-rays (that is electromagnetic waves of high frequency) impinge on a metal, the metal becomes charged because something endowed with negative charge is emitted. The phenomenon was called the photoelectric effect. The first observation can be traced back to Alexandre Becquerel (1839), followed, more extensively, by Heinrich Rudolf Hertz (1887), Wilhelm Hallwachs (1888), Augusto Righi, Aleksandr Grigorievich Stoletov, Julius Elster, and Hans Friedrich Geitel (from 1889 to 1902). In 1899, Joseph Thomson inferred that the so-called cathode rays were particles with negative charges. Later, such particles were called *electrons*, one of the main components of any atom. From 1899 to 1902, Philipp von Lenard investigated how light frequency affects the energy of the electrons released by the photoelectric effect. Experimental data accumulated, but no satisfactory theoretical explanation was found. Albert Einstein explained this effect in 1905. In the years that followed, heated discussions occurred among scientists about the theory of light and the quanta. The wave–particle duality permeates not only considerations on light, but also studies of elementary particles, atoms, and molecules. Other famous scientists contributed to this effort in the 1920s and the 1930s. Even though some of them were reluctant to accept or were opposed to the new ideas, the path toward quantum mechanics and quantum electrodynamics (which includes Einstein's special relativity) was unavoidable. In particular, the quantum and wave aspects both confirm that light can exert pressure on a material surface. In addition, even when light impinges on such a surface, both the principles of energy and momentum conservation still hold. These facts are quite important for space sailing.

Benefits for Spaceflight

Can the above property of photons be utilized for space propulsion? In the first years of the 20th century, Swedish chemist Svante August Arrhenius suggested that spores, pushed by solar-light pressure, might diffuse life through the solar system and beyond. In the 1920s, Russian scientists Konstantin Tsiolkovsky and Friedrich Arturowitsch Zander wrote that a very thin space sheet, thrusted by solar-light pressure, should be able to achieve high speeds in space. Subsequently, the Americans Carl Wiley (1951) and Richard Garwin (1958) published the first technical papers on solar sails of the modern era. Garwin was the first to use the term *solar sailing* applied to space vehicles. Other scientific papers were published by T.C. Tsu (1959), H.S. London (1960), N. Sands (1961), and W.R. Fimple (1962); they regarded solar-sail spacecraft trajectories that are different from the usual Keplerian orbits.

An intense flux of sunlight impinges on any material surface in space. However, in general, the correct force of the solar photons on a surface cannot be obtained by simply multiplying the solar photon pressure (see Glossary) by the surface's area. It is the twofold photon nature that intervenes in the interaction between light and material surface. The main aspects of this important interaction are presented qualitatively in Part II, whereas a more technical description can be found in Part IV of this book. Here is an example of an ideal case. If sunlight is fully reflected by a surface, then the force on a perpendicular sail of 0.11 km^2 at 1 AU would amount to 1 newton (0.225 lbf). If the whole sailcraft has a mass of 169 kg (372.6 lb), then the solar-pressure acceleration on this space vehicle equals the Sun's gravitational acceleration at 1 AU or 0.00593 m/s^2. (This is the same acceleration that compels Earth to revolve about the Sun). It is very challenging to design a spacecraft with a mass-to-area ratio equal to 169 kg / 0.11 km^2 = 1.53 g/m^2.

In the following years, there was a certain interest in such options for navigating in space, but most of the attention by propulsion and mission designers was focused on energetic chemical engines, advanced rocket engines such as the full nuclear rocket, and the nuclear-electric thruster. In the 1970s, the NASA Jet Propulsion Laboratory studied a solar-sail interplanetary probe for a rendezvous with Comet Halley (which was returning to the inner solar system in 1986), but the study did not lead to a flight article. In the 1970s and 1980s, extra-solar and interstellar space sails pushed by photons were proposed and analyzed deeply by G.L. Matloff, E.F. Mallove, M. Meot-Ner, and R.L. Forward. Many new mission concepts and strategies were introduced. Solar sails moving in the solar system were analyzed also by J.C. van der Ha, V.J. Modi, R.L. Staehle, and E. Polyakhova. Such studies helped to develop understanding of the potentialities of the pressure of light—together with appropriate spacecraft technology—for many types of space mission, from those ones in the Earth–Moon space to missions to the near interstellar medium and beyond.

In the 1990s and in the first years of this century, new meaningful studies, findings, and technological realizations have been made in the U.S., Europe, and Japan. After NASA and the Japanese space agency (JAXA), ESA and other national European space agencies have become aware—through a number of meaningful (either independent or contract) studies—of the importance of the solar-sail option for advanced space exploration and utilization. The main consequences of such historical and recent findings are the subject of Part II of this book.

We conclude by emphasizing some key points peculiar to space solar sailing:

- Since the solar photon flow is constantly present in the solar system, a spacecraft equipped with a sail undergoes a continuous thrust that allows it to reach any destination,
- There is no need for the expenditure of on-board propellant during interplanetary travel,
- As is the case with terrestrial sailboats, space sailcraft require some device that can control the orientation of the sail,
- As is the case with terrestrial sailboats, interplanetary sailcraft can be fully reusable,
- There is no need to build a staged sailcraft (unless one wants to accomplish some very special missions),
- Also depending on how good we are in navigating, flight times for certain missions can be decreased significantly with respect to those obtainable by rockets.

There are other properties relevant to space sailcraft; some of them will be explained in Part II and, for the interested, mathematically inclined reader, in Part IV.

Further Reading

Phoenicians

http://phoenicia.org/index.shtml.
http://phoenicia.org/cities.html.
http://www.worldwideschool.org/library/books/hst/ancient/HistoryofPhoe-nicia.
http://en.wikipedia.org/wiki/Phoenicia.
http://www.cedarland.org/ships.html
http://phoenicia.org/ships.html.
http://www.barca.fsnet.co.uk/basics.htm.
http://www.mnsu.edu/emuseum/prehistory/aegean/theculturesofgreece/phoenician.html.
http://www.lgic.org/en/phoenicians.php.
http://www.bestofsicily.com/mag/art150.htm.
http://www.cedarseed.com/water/phoenicians.html.

Sea Sailing

Bob Bond, *The Handbook of Sailing*, Knopf, New York, 1992.

Ross Garrett, Dave Wilkie, *The Symmetry of Sailing: The Physics of Sailing for Yachtsmen*, Sheridan House, 1996.

J. J. Isler, Peter Isler, *Sailing for Dummies*, Hungry Minds, Inc. 1997.

Bryon D. Anderson, *The Physics of Sailing Explained*, Sheridan House, 2003.

http://www.nationalgeographic.com/volvooceanrace/interactives/sailing/index.html.

http://www.wb-sails.fi/news/SailPowerCalc/SailPowerCalc.htm.

History of Physics and Light Pressure

R. D. Purrington, *Physics in the Nineteenth Century*, Rutgers University Press, 1997.

http://www.aip.org/history/einstein/.

http://en.wikipedia.org/wiki/History_of_physics.

http://www.biocrawler.com/encyclopedia/Radiation_pressure.

http://galileo.phys.virginia.edu/classes/252/home.html.

http://www.britannica.com/eb/article-9062400.

Space Missions by Sail

Principles of Space Sailing

6

The romantic-sounding term *solar sail* evokes an image of a majestic vessel (similar to the great sailing ships of the 18th century) cruising the depths of interplanetary space (Fig. 6.1). In a very literal sense, this imagery is very close to the anticipated reality of solar sails. Very large and diaphanous sail-propelled ships will traverse our solar system and perhaps, one day in the future, voyage to another star. From what will these ships be made and how will they work?

The solar wind is a stream of charged particles (mostly hydrogen and helium) emitted by the Sun. The solar sails, which are the primary focus of this book, are not blown by the solar wind, though there have been proposed "sails" that will do just that. The "wind" that blows a solar sail is sunlight. The ever-present, gentle push of sunlight will eventually accelerate our

Figure 6.1. Solar sails will propel our starships in much the same way that wind gave the great sailing ships their energy for more earthly exploration. (Courtesy of NASA) (See also color insert.)

G. Vulpetti et al., *Solar Sails*, DOI: 10.1007/978-0-387-68500-7_6,

starships to speeds far above that achievable by chemical or electric rockets—or the solar wind.

What Is a Solar Sail?

To understand how sunlight propels a solar sail, one must first understand at least a little bit about the interaction of light with matter. When sunlight, which has momentum, falls on an absorptive surface (consider a surface painted black), very little sunlight reflects from the surface; most is absorbed. In space, where there is no air resistance and an object is essentially free from other forces, the sunlight falling on a black sail will transfer its momentum to the sail, causing the sail to move. If the same material is now painted with a light-reflecting material (like a mirror), it will reflect the photon instead of absorbing it. Like the black sail, this one will also begin to move, and the reflective sail will accelerate at a higher rate than the one with a dark surface. The reflected light transmits more of its momentum to the sail than the light that was absorbed. The principle of momentum transfer applies to all forms of sails, including photon sails, magnetic sails, plasma sails, and, very recently, electric sails.

Momentum Transfer

You can test this at home using a rubber ball, a ball made of modeling clay (or Play Doh™), and a hinged door. First, throw the ball of modeling clay at an open door and notice how far the door moves. The clay will most likely stick to the door, mimicking the absorption of light on a dark-colored sail. Next, open the door back to its initial position and throw the rubber ball, trying to throw with the same force as was used with the clay, and notice how far the door moves. If the experiment goes as it should (which is not always the case in experimental physics!), the rubber ball will bounce off the door and cause it to close farther than was achieved with the ball made of clay. In this case, the rubber ball (like the light) is reflected from the door, transferring twice as much momentum to the door as the ball of clay. This is analogous to the light reflecting from the sail.

At first thought, it might appear that a solar sail would be very limited in the directions it can move. For example, it seems intuitive that a solar sail might be used for a voyage to Mars or Jupiter, but not to Venus or Mercury. Venus and Mercury are sunward of Earth and one might think that the Sun will therefore constantly push a sail away from them. If the planets were not in orbit about the Sun, this would be correct. But the planets *are* in orbit the Sun and we can take advantage of this fact to allow a solar sail to fly either toward or away from it.

Just like a wind-powered sailing ship, a solar sail can tack—sort of. Instead of maneuvering back and forth "into" a head wind so as to move the ship toward the prevailing wind (sunlight), the sail can be tilted to alter the angle at which the light strikes and reflects from the sail—causing it to either accelerate or decelerate. Earth orbits the Sun at 30 kilometers per second (>66,000 miles per hour), and any sail launched into space from Earth will therefore be in an orbit around the Sun with about the same orbital velocity. Since the distance a planet or spacecraft orbits around the Sun is determined by how fast it is moving, one may change that distance by either speeding up or slowing down. For a solar sail, this means changing the orientation of the sail so that it reflects light at an angle such that the momentum from the sunlight pushes the sail either in the direction it is already moving (acceleration) or in the opposite direction (deceleration). In either case, part of the light's momentum will be perpendicular to the direction of motion, causing the sailcraft to move slightly outward at the same time it is accelerating or decelerating. Adding up the various forces can be complicated, and making sure the net force is causing motion in the desired direction is an engineering challenge. Fortunately, we know how to model these effects and control them, just like a seasoned captain knows how to tack his boat against the prevailing wind.

Steamships and modern diesel-electric cruise ships must refuel or they will be dead in the water. As long as the wind blows, a sailboat will be able to move. Like steamships, rockets must refuel. Solar sail craft needn't bother! As long as the Sun shines, they will be able to use the sunlight to move. Unfortunately, this means they can only accelerate or decelerate in the inner solar system where sunlight is plentiful. When they reach distant Jupiter, the available sunlight is only a fraction of that available on Earth and the resulting forces on the sail are too weak. As we will discuss in later chapters, there are tricks that may be used to allow a solar sail to traverse the entire solar system and perhaps take us to the stars.

In order to work, a solar sail must be of very low mass. The momentum transferred from sunlight to the sail is very small. If the sail and its payload are massive, the resulting acceleration will be slight. Simply stated, heavy is bad. What is needed are highly reflective, strong, and lightweight sails. Modern materials science has provided several promising candidates and building viable sails from them is now within our reach.

How Can the Solar Wind Be Used for Sailing?

As mentioned above, there are other sail concepts that use entirely different physical processes to sail through space. Since three of them use the solar wind, it will be useful to discuss the nature of that "wind" before describing how they harness it to produce thrust.

The solar wind is an ensemble of electrons and positively charged ions (mostly hydrogen and helium) produced by the Sun. Just like sunlight, there is a continuous stream of this plasma flowing outward from the Sun into the solar system. Unlike sunlight, there may be intense bursts of these charged particles emitted by the Sun at any time and in any direction. These ions and electrons race outward from the Sun at speeds in excess of 400 kilometers per second. In fact, during periods of high sunspot activity, these speeds have been measured to be greater than 800 kilometers per second! Could we use this wind to propel our spaceships?

One way to take advantage of the solar wind for propulsion is the Magsail. As the name implies, a magsail uses the interaction of the solar wind with a magnetic field to produce thrust. A charged particle moving through or into a magnetic field will experience a force, causing it to speed up, slow down, or change direction, depending on the direction in which it is moving with respect to the field. And since Newton taught us that "for every action there is an equal and opposite reaction," the magnetic field will likewise be affected. In this case, the structure from which the field originates will experience the opposing force, giving it acceleration.

Conventional magnets made of iron are heavy. After all, they are made of iron. Flowing a current through a wire can make lighter weight magnets. Flowing a large current in a low-resistance wire will produce a strong magnetic field. Magsail designers postulate the use of large superconducting wire loops carrying high currents to interact with the solar wind—sailing the solar wind.

While technically interesting and somewhat elegant, magsails have significant disadvantages when compared to solar sails. First of all, we don't (yet) have the materials required to build them. Second, the solar wind is neither constant nor uniform. Combining the spurious nature of the solar wind flux with the fact that controlled reflection of solar wind ions is a technique we have not yet mastered, the notion of sailing in this manner becomes akin to tossing a message in a bottle into the surf at high tide, hoping the currents will carry the bottle to where you want it to go.

A cousin of the magsail is the plasma sail. Like the magsail, a plasma sail would use the solar wind for propulsion. Instead of interacting with the magnetic field produced by a large superconducting magnet, however, the

Figure 6.2. The plasma sail would use the solar wind to propel it outward into the solar system. (Courtesy of R.H. Winglee) (See also color insert.)

plasma sail would derive its thrust from a bubble of plasma surrounding the spacecraft. This plasma in this bubble would be pushed by the solar wind, dragging the magnetic field in which it is trapped, and, consequently, the spacecraft that produced it (Fig. 6.2).

A fundamental aspect of electromagnetism is the fact that like charges repel each other and opposite charges attract. Electrons, which carry a negative electrical charge, will repel one another. Similarly, two positively charged ions would also repel each other. Another aspect of charged particles is that their motion is affected by a magnetic field. As discussed above, a charged particle moving though a magnetic field will experience a force acting upon it. If the field is properly aligned and of sufficient strength, then the charged particles will be trapped, forever moving along the field line, spiraling along it until some external force moves them away.

The plasma sail would use electromagnets to generate a magnetic field. A plasma, which is a gas composed of electrons, ions, and their electromagnetic forces, would then be injected into the field to form a plasma "bubble" around the magnet (and the spaceship carrying it). This plasma bubble could theoretically inflate to more than 60 kilometers across. As the solar wind impinges the bubble, it will be forced to move around it by the interactions of the charged particles in the solar wind and those in the plasma bubble. The net result is that the solar wind will push the bubble forward, much like you can blow a balloon outward from your hand with a gentle exhale.

Unfortunately, like the magsail, the forces acting on the plasma sail would make the direction in which the spacecraft moves hard to control. It also requires extremely powerful and lightweight magnets, which we do not yet have the capability to build. And then there is the concern that in the real world, the solar wind might just rip the plasma bubble away from the spacecraft, leaving it stranded with no sail whatsoever. Until it is tested in space, we will not be able to verify that it will work at all.

In 2004, a new concept arose for trying to utilize the momentum flux of solar wind: the electric sail. Similarly to the magsail, this concept uses the solar wind for producing thrust. However, differently from the magsail, this sail interacts with the solar plasma via a mesh of long and thin tethers kept at high positive voltage by means of an onboard electron gun. In its baseline configuration, the spacecraft spins and the tethers are tensioned by centrifugal acceleration. It should be possible to control each wire voltage singly, at least to within certain limits. Thrust originates since the solar-wind protons (remember that any proton in the Universe is positively charged) are repelled by the positive voltage of the mesh. In contrast, the electrons are first captured and then ejected away by an onboard electron emitter because accumulation of electrons would neutralize the mesh voltage rapidly. (The reverse configuration, i.e., electrons repelled and protons re-emitted, would produce a thrust about 2000 times lower). Figure 6.3 is a sketch of the electric-sail concept. At this point you can object that the solar-wind fluctuations are always present and no trajectory design would be reliable, quite analogously to the magnetic sails. However, this spacecraft could control directly the electric field that fills the space around it. In particular, the magnitude of the thrust could be controlled between zero and some maximum value by adjusting the electron gun current or voltage. Are such advantages sufficient, for instance, to issue a thrust level almost independent of the high variable solar-wind intensity? As of 2008, the answer is not known and research is in progress just to study this basic aspect.

It is important to realize that any propulsion type needs to be controlled for designing the vehicle's motion with high probability (the mathematical certainty is not achievable in practice). Otherwise one could not know where it is going to or when it arrives at the target. Solar sailing cannot be an exception. Even sunlight is variable with time and mostly unpredictable. However, the fluctuation level is very low and we can design/predict a

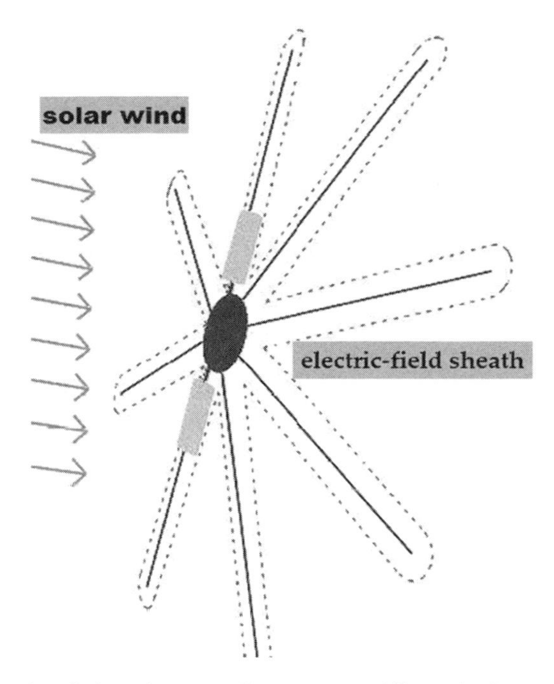

Figure 6.3. Sketch of the electric-sail concept. Differently from the magnetic-sail mode, current and voltage of the onboard field generator could be controlled. (Courtesy of Dr. Pekka Janhunen)

mission in all phases. Perhaps, this is the biggest difference between sunlight-based and solar–wind based sailcraft.

Solar sails, magsails, plasma sails, and electric sails are all examples of the creativity of the human mind unleashed. Using the immense energy of the Sun for propulsion is an idea whose time has come, and solar sails are poised to be the first to make use of this never-ending supply of fuel for space exploration.

Further Reading

For readers interested in science-fiction stories that use the solar sail as a primary means of propulsion, we recommend Arthur C. Clarke, *Project Solar Sail*, ROC/Penguin, New York, 1990. A more technical treatment of solar sails can be found in Louis Friedman, *Starsailing, Solar Sails and Interstellar Travel*, Wiley, New York, 1988.

What Is a Space Sailcraft? 7

Chapters 1 to 4 discussed the importance of the rocket propulsion in the first 50 years of spaceflight, and its limitations with respect to what space-faring nations (augmenting in number and quality) would want to accomplish in the solar system and beyond. Chapter 5 discussed the concept of sailing, first on Earth seas with conventional sailboats, then by extending the concept to space; there, the first similarities and differences between sea sailcraft and space sailcraft were emphasized. Chapter 6 detailed the principles of space sailing. Now we discuss what a space sail actually means through the great impact it can have on the design of the different systems, which is not as obvious as it might seem.

One may think of the space sailcraft as the sum of two pieces: something like a conventional spacecraft (containing the payload) and a sail system consisting of a sail with mast, spars, rigging, tendons, and a device controlling its orientation in space. That's correct, in principle. However, such an oversimplified description may induce someone to believe that building a sailcraft means merely adding a sail to something that already is well known. In Chapter 5, we mentioned basic analogies and differences between terrestrial sailboats and space sailcraft. Here there is another important difference—the relative size: *In the space sailcraft, the two-dimensional size of the sail system overwhelms that of any other system.* This is due to three reasons: (1) Earth orbits the Sun at about 1 astronomical unit (AU). (2) The Sun's power emitted from its surface (technically called the solar radiant emittance or exitance) amounts to about 63.1 million watts per square meter. (3) The linear momentum a photon transports is scaled by the factor 1/c, where c denotes the speed of light in vacuum. As a result, an object of one square meter that is 1 AU distant from the Sun and perpendicular to the sunlight's direction can receive about 1366 watts (on average during a solar cycle). What does it mean? If this object were a perfect mirror, it would experience a force equal to 2*1366/c = 0.0000091 newtons (or about 0.000002 lbf). If the mass of such a body were 91 grams, the ensuing acceleration would amount to 0.1 mm/s^2. At 1 AU again, the solar

G. Vulpetti et al., *Solar Sails*, DOI: 10.1007/978-0-387-68500-7_7,

gravitational acceleration (which allows Earth to orbit stably about the Sun) is 5.93 mm/s^2. In other words, the solar-light pressure acceleration on this particular object would be about 1/60 of the solar gravity at 1 AU.

The previous example (a typical one in solar-sailing books) tells us two important things. First, such an acceleration level would be sufficient for many space missions (especially the first ones) and would correspond to an object having a mass-to-area ratio of 91 g/m^2; second, if we aim at ambitious missions, we have to lessen this ratio by a factor of ten, at least. Despite the significant advancement in materials technology, key space systems (including the whole sail system) cannot be designed by decreasing their mass arbitrarily. As a result, a sailcraft has to have a large sail, from a few thousands to many ten thousands of square meters to begin with. (At the end of this chapter, we will discuss the micro-sailcraft concept).

Now that we understand the above statement about relative size, we can analyze some implications of the major spacecraft systems. In this chapter, we adopt the following nomenclature: Sailcraft = Sail System + Spacecraft. Some of the topics briefly discussed here complement the discussion in Chapters 11 and 12.

Sail Deployment

Normally, once the whole sail system is manufactured on the ground, it should be folded and placed in a box. Subsequently, it will be unfolded in an initial orbit and, then, some initial orientation will be acquired. It is easy to guess that the sail system is considerably delicate. The sail configuration and the related deployment method affect the performance of the solar-sail thrust, which is still a work in progress; some 20-m by 20-m sails have been unfolded in important experiments on the ground. This research area is considerably broad, and any deployment method must pass future tests in space. Let us mention just a few issues related to sail performance. Suppose that the sail is unfolded by means of telescopic booms, which slowly come out of the box. This means that the sail, either squared or polygonal in shape, has been divided into smaller (e.g., triangular) sheets. These sheets could be considered as a membrane subjected to two-dimensional different tensions in their plane. If the sheet undergoes a tension that it is much lower than the other-dimension tension, then wrinkles develop. However, a sail divided into parts presents advantages from the construction and handling viewpoint.

Wrinkles should be avoided as much as possible because multiple reflections of light can occur among them. These wrinkles cause two undesirable effects: (1) locally, the sail can absorb much more energy than it would in normal conditions, and so-called hot spots develop; (2) if wrinkles cover a large fraction of the sail, the solar-pressure thrust decreases with respect to what is expected for a flat smooth surface. In one view, wrinkles increase the sail's intrinsic roughness (coming from the sail manufacturing process), which lessens the surface's ability to reflect the light in a specular way.

Other deployment methods, some of which have been tested on the ground, apply to circular sails. For instance, the sail would be unfolded by a small-diameter inflatable tube attached around the sail circumference; once deployed, the tube has to be rigidized (in the space environment) to retain its shape without the need of keeping the tube under pressure (a thing impossible to do for a long time). Although some corrugation may arise from such a method, it is expected that the sail could be almost wrinkle-free. One should note, however, that, replacing telescopic booms by inflatable tubes does not avoid wrinkles; the important thing is the circumferential geometry of the supporting beam (see the discussion of the Aurora collaboration in Chapter 13).

Sail Control

This topic is discussed in detail in Chapter 11. Here, we stress just a few issues that characterize a sailcraft from our viewpoint. After the separation of the *packed* sailcraft from the launcher, the first maneuver, the related commands and procedures (the so-called attitude acquisition) are performed in order to begin the planned mission time sequence. The first part of the sequence includes sail deployment. After sail unfolding and checkout (e.g., via the television cameras of the sail monitoring system) have been completed, the sail has to be oriented stably toward the Sun (not necessarily normal to the sunlight). The sail's first orientation maneuver (which can be considered the second *attitude acquisition*) is probably accomplished via some traditional equipment such as cold-gas thrusters, rotating wheels, and extendable booms. Other ways can be developed. When the solar photons impinge on the sail, the center of pressure rises, as the sea wind does when it swells the sails of a conventional sailboat (see Chapter 5). From that moment on, two objects—the spacecraft and the sail system—are both subjected to gravity, and will move through the action of the sail on the spacecraft and the reaction of the spacecraft upon the sail. However, since the spacecraft and sail do not form a rigid body, it should be possible to accomplish relative movement between the center of mass (of the sailcraft) and the center of pressure (of the sail). (This operation will involve only small electric motors.) The result will be a change in the sail orientation. Whereas Chapter 11 focuses on sail attitude control, here we note that small mass variations of the sailcraft cannot be excluded in a mission, although heavy amounts of propellant should be avoided because the primary propulsion comes from solar energy.

Communication System

Let us consider the communication between the sailcraft and the ground station(s). Communications between the spacecraft and the ground control center are fundamental in a space mission, but the control center is not the only base. The spacecraft has to be tracked periodically from other ground stations with different tasks. NASA's Deep Space Network, ESA's set of ground stations, and national centers (from different countries) are examples of ground stations. Both stations and the control center receive and send electromagnetic waves from and to the spacecraft in different frequency bands. To do so, the spacecraft has to be "electromagnetically visible," and the onboard antennas have to point to Earth. Here is another implication of relative size. Where do we allocate the onboard antennas? This depends not only on the sail configuration, but also on the sail orientation along the sailcraft trajectory. On a spacecraft, there may be different types of antennas: scientific-data-return high-gain antennas, telemetry and command antennas, emergency and low-gain antennas. Normally, a high-gain dish antenna works in different bands and thus performs different functions. Although it is very thin, the sail can cause obstruction of the antenna waves. It would not be very wise to put antennas close to the sail rim, as it could (1) cause mechanical and electrical problems, (2) induce sail instability, and (3) make the normal sail control much more difficult. A possible solution may be to use the structure that normally forms the "axis" of the sail; for each antenna type necessary for the mission, we can place one on the front side of the sail and (a copy of) this one on the back side. In future advanced missions beyond the solar system, a small part of a wide sail might be designed to function as a big antenna, so large amounts of scientific data may be downloaded to Earth-based or Moon-based receiving antennas from distances as large as hundreds of astronomical units.

Sailcraft Temperature

As for the power system on board a sailcraft, it is obvious that the required amount of watts depends on the mission type and purposes. The power system has to supply energy also to the thermal-control system. Space vehicles have to be designed to withstand the temperatures of space environments. Sail temperature can be adjusted solely by changing its orientation with respect to the incident light, but not too much, otherwise the sailcraft trajectory would change considerably. One has to design a trajectory by satisfying the temperature requirements of the sail materials and the mission target(s). In addition, the nonreflected photon energy is absorbed by the sail and then re-emitted almost uniformly. Therefore, if the sailcraft is sufficiently close to the Sun, other spacecraft systems may be hit not only by part of the light diffused by the sail, but also by a significant amount of energy in the form of infrared radiation, almost independently of their positions with respect to the sail. Therefore, the thermal control of such systems requires additional power in order to keep their range of operational temperatures.

A different situation occurs from sailcraft entering planetary shadows (penumbra and umbra). Since the sail is extended and very thin, the sail temperature immediately drops and adapts to the space environment. When the sailcraft returns to light, the sail temperature rises much more quickly. Although the space environment around a planet is very different from the interstellar medium, the sail's temperature jumps may achieve almost 200 K (in some missions). Therefore, the sail materials have to be selected to withstand many high–low–high temperature cycles during their years of operational life.

Payload

Usually, the mission payload consists of a set of instruments for detecting particles and fields, for receiving and sending signals, for taking pictures of objects, and so on. Can the payload be affected by the sail? Suppose we design a planetocentric sailcraft, the payload of which will measure the detailed structure of the planet's magnetic field, if any, in a large volume around the planet itself. The solar wind, interacting with such a magnetic region, continuously changes its shape and properties. One of the first sailcraft missions will probably be of a similar kind, for which the planet is Earth. Incidentally, although space satellites such as the NASA IMAGE spacecraft (March 2000–December 2005) and the ESA four-satellite

CLUSTER (in operation since August 2000) have discovered fundamental phenomena in Earth's magnetosphere, nevertheless there are many physical quantities to be measured better and longer in our magnetosphere.

How does a sailcraft behave inside a large region of magnetic and electric fields, and with many flows of charged particles? (Earth's magnetosphere does not protect the planet completely.) The sail size is wider than characteristic plasma lengths; one of the expected effects consists of space plasma surrounding the sail's front side by a positively charged sheath, whereas a wake of negatively charged flow extends beyond the sail's back side significantly. Such a charge distribution changes the local properties of what the payload instruments can measure. Therefore, it is important to locate the scientific sensors sufficiently *ahead* of the sail system, where the plasma will be *undisturbed* by the sailcraft's presence.

Since each mission has its own features, the payload-sail arrangement should be analyzed on a case-by-case basis.

The Micro-Sailcraft Concept

In Chapter 12, we will discuss nanotechnology and its potential impact on solar sails. Here, we limit the discussion to the following questions. Were the ratio between the sailcraft mass and its effective area kept fixed with the same sail orientation, would the motion of the vehicle remain unchanged regardless of the sail size? What would happen if the sailcraft were scaled down further? In other words, how much can we reduce the size of the sailcraft? Is it only a technological problem or is there any physical limit that prevents having an (almost) arbitrarily small vehicle?

Let us start by noting that about 98 percent of the solar irradiance is due to photons with wavelengths from 0.25 microns to 3.5 microns (micrometers). The visible part of the spectrum (from 0.4 to 0.7 μm) carries about 37 percent of the total solar irradiance. If one wants to utilize the solar energy at its best, it is difficult to think of building a sail with a diameter less than 10 microns. Thus far, the telecom system has been based mainly on microwaves. Even if one envisages a complete system transmitting information at 100 GHz, the only antenna could not be smaller than 3 millimeters. If one turns to telecom system via laser, small lasers are possible, but there are other problems (e.g., pointing accuracy and receiving ground telescope) to be taken into account.

Consider a scientific payload. Interstellar spacecraft of 1 kg have been proposed; however, if one wishes to accomplish some high-performance deep-space mission science by tiny volume detectors, the probability of interaction between any space particle and the detector decreases dramatically. Even if we have one-event (large) detectors, getting a sufficiently high number of events is fundamental for analyzing data the mission is seeking. The minimum size of scientific instruments can vary significantly; it depends not only on technology, but also on the underlying physics.

What about nanoscience and nanotechnology for solar sailing? These quite intriguing topics deserve attention; they will be discussed in Chapter 12 with regard to sailcraft. Here, we note that a few years ago, author Matloff discussed the possibility of a swarm of many tiny spacecraft, or nanoprobes, that collectively behave like a large spacecraft. As the reader can see, this is a very advanced concept; in principle, the probe might look like many small antennas that act together as a very large nonconstructible antenna, but much more intricatc. This concept will be analyzed more deeply as nanoscience develops.

Conclusion

Most of the above-mentioned problems can be solved, as many other problems have been in the history of spaceflight. This chapter has shown that the sailcraft represents something considerably different from the satellites and probes launched into space so far, and we are just at the beginning.

Figure 3.4. An ion thruster on the test stand. (Courtesy of NASA)

Figure 4.1. A rigid aeroshell could protect a payload during aerocapture. (Courtesy of NASA)

Figure 4.3. A satellite moves toward a higher orbit after release from a tether-equipped spacecraft. (Courtesy of NASA)

Figure 4.4. The solar wind's interaction with Earth's magnetosphere. (Courtesy of NASA)

Figure 6.1. Solar sails will propel our starships in much the same way that wind gave the great sailing ships their energy for more earthly exploration. (Courtesy of NASA)

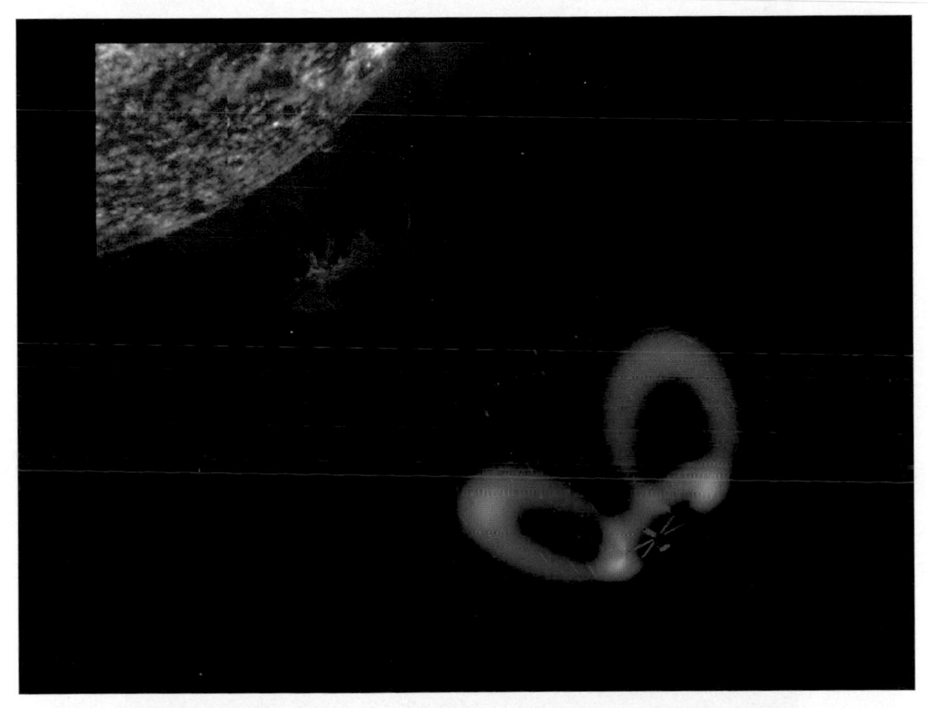

Figure 6.2. The plasma sail would use the solar wind to propel it outward into the solar system. (Courtesy of R.H. Winglee)

Figure 8.3. Ariane 501, the maiden flight of Europe's Ariane 5 launcher, veers off course and is destroyed by its automated destruct system. (Courtesy of ESA)

Figure 8.4. A test of the Space Shuttle's main engines. (Courtesy of NASA)

Figure 11.4. The Russians were the first to deploy a spinning, solar sail–like structure in space. This is an artist's concept of Znamya-2. (Courtesy of Russian Space Agency)

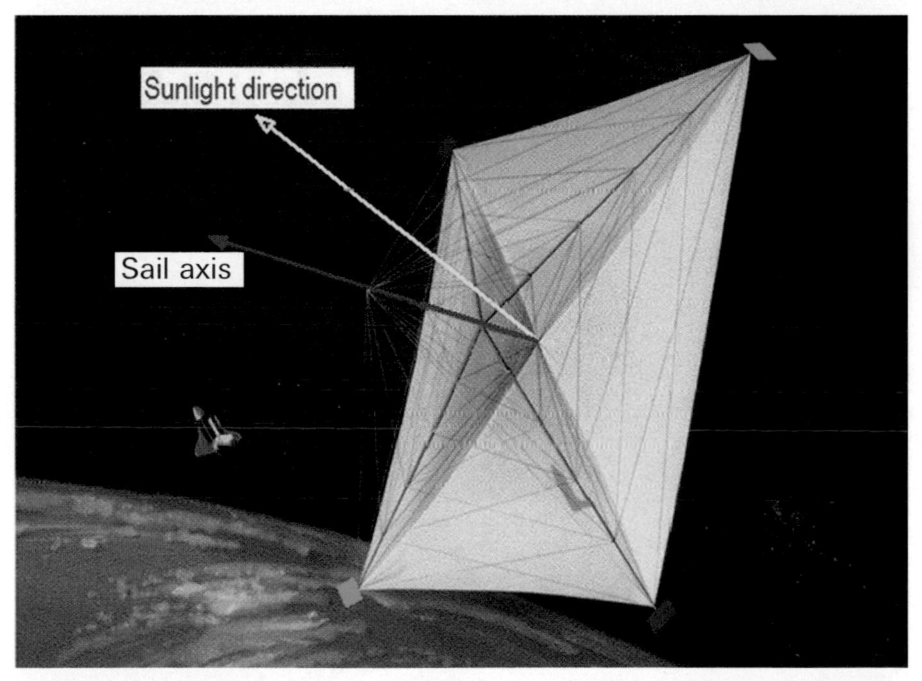

Figure 11.6. Solar sail controlled in attitude by small sails located on the boom-tips. (Courtesy of NASA, adapted by G. Vulpetti)

Figure 12.1. A photograph of the DLR solar sail ground demonstrator under full deployment. (Courtesy of DLR)

Figure 12.2. Sketch of The Planetary Society's Cosmos 1. (Courtesy of The Planetary Society.)

Figure 12.3. The L'Garde solar sail before vacuum testing at NASA's Plum Brook facility. The scale of the device can be appreciated by examining the relative size of the people in front of it. (Courtesy of NASA)

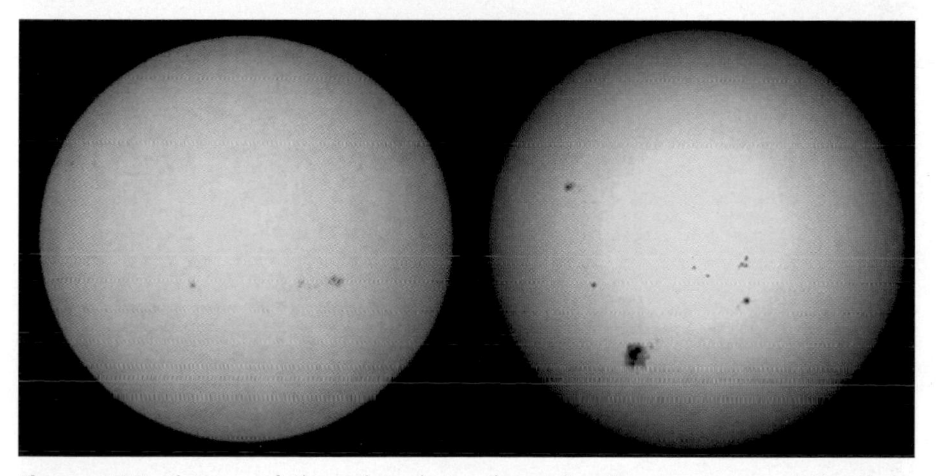

Figure 15.7. Pictures of the solar photosphere showing sunspots and the limb-darkening effect. (Courtesy of NASA)

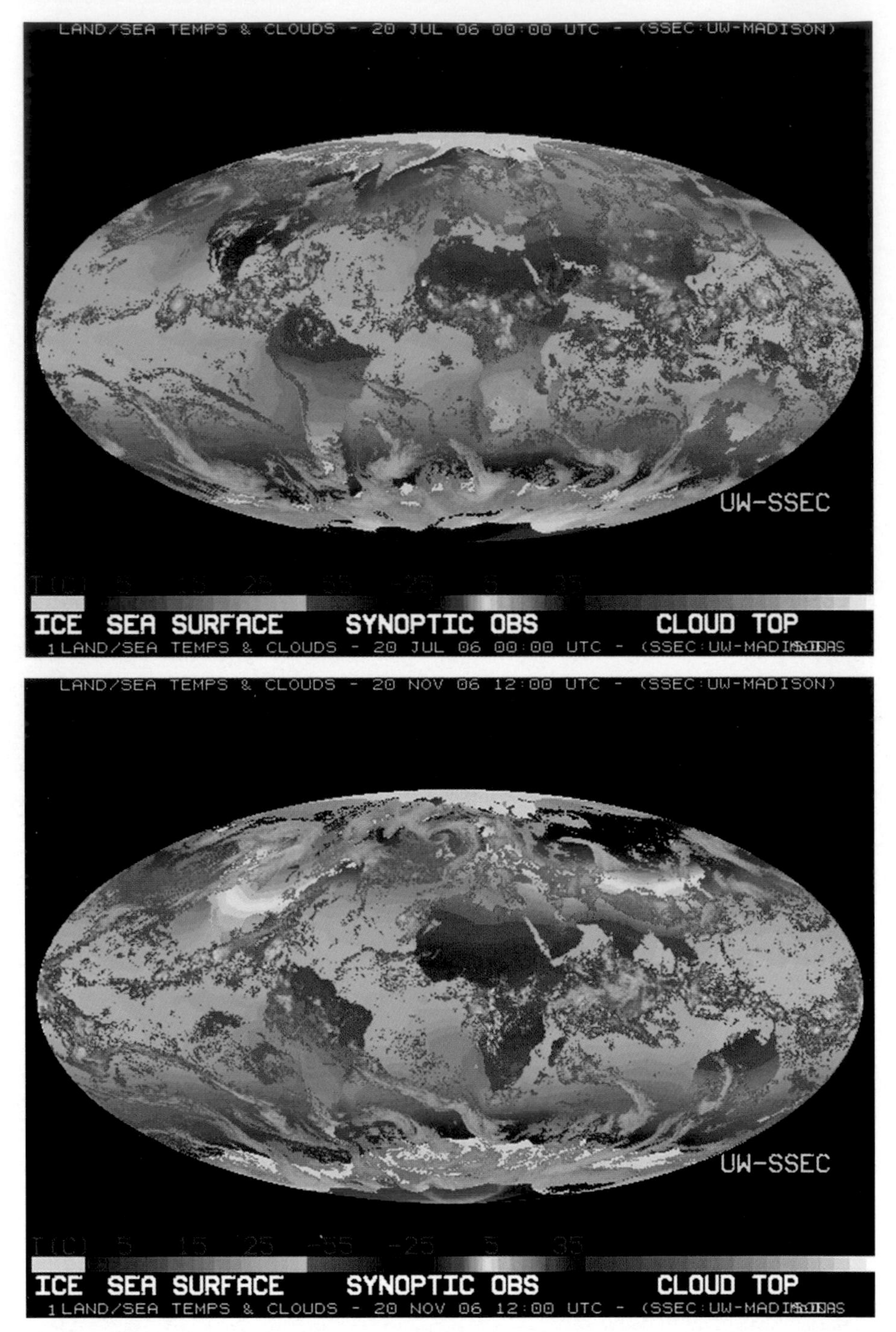

Figure 15.8. Globe geostationary satellite composite images for 2 days in 2006. (Courtesy University of Wisconsin at Madison)

Sails Versus Rockets

8

Rockets move spacecraft around in space from one destination to another. Solar sails also move spacecraft around in space from one destination to another. That is just about the only similarity between these two methods of spacecraft propulsion—commonality of function. Once you get to the next level and begin to describe how they work, their processes and support systems, and the mission-level requirements they each possess, the similarity ends—with a vengeance. In this chapter, characteristics peculiar to solar sails are in italics.

Rockets and Bombs

A rocket is essentially a bomb that goes off slowly. As was tragically seen in the loss of the American Space Shuttle Challenger and unmanned rockets too numerous to count, a rocket failure produces a spectacular explosion. This failure is often caused by nothing more than the rocket system's inability to contain the chemical reaction producing its energy.

In a conventional rocket, combinations of chemicals are mixed together in a manner so as to produce energy, resulting in the rapid expansion of a hot gas. If properly contained and directed, this hot gas safely and in a controlled manner exits the rocket producing thrust. If it is not properly contained, then the hot gas rapidly expands outward from the combustion chamber in more than one direction—having the same net effect as a bomb. If the reaction products are not appropriately directed, then the rocket cannot produce thrust in a useful manner, while not as spectacular a failure as an explosion, having the rocket fly off in an unplanned direction is as much of a mission failure as if it had exploded.

Having mentioned Challenger, it might be instructive to look at the "bomb equivalent" result that occurs when the energy expended during a rocket launch is instead used to blow things up. In a rocket, a fuel is mixed with an oxidizer, resulting in an energy-producing reaction that rapidly

G. Vulpetti et al., *Solar Sails*, DOI: 10.1007/978-0-387-68500-7_8,

heats a gas, causing it to expand through a nozzle in a direction opposite to the direction of desired motion. This is a fancy way of saying that we burn fuel in a tank with an opening on one end. Burn the fuel in the same tank—this time without an opening—and you get an explosion as the pressure resulting from the combustion increases to the point where the tank walls fail. This is a bomb.

As many a rocket designer will tell you, getting the bomb to go off slowly is not an easy job. The history of almost all successful rockets begins with the designers learning this hard lesson. Here are but a few examples:

1. Leading up to the launch of America's first satellite was a string of impressive launch failures. Shown in Figure 8.1 is the U. S. Navy's Vanguard Rocket exploding at launch on December 6, 1957.
2. The N-1 was the rocket that would have sent Soviet cosmonauts to the moon, potentially beating the Americans there. But it was not to be so. Between 1969 and 1972 all 4 attempts to launch the N-1 ended in failure. The launch attempt made just weeks before the successful liftoff of Apollo 11 ended in a disaster that some consider to be the biggest explosion in the history of rocketry. Figure 8.2 shows two N-1 rockets on their launch pads in the Soviet Union.
3. The Europeans have also suffered their share of launch failures. The most spectacular was probably that of the first launch of the Ariane-5 rocket. In June 1996, the inaugural flight of the Ariane 5 was to carry a set of science spacecraft into space. Instead, it blew them apart. Figure 8.3 is a photograph of what happened to that ill-fated rocket at launch, just 40 seconds into flight.
4. Not all rocket failures happen on the ground. Catastrophic failures happen in deep space as well. For example, the Mars Observer mission, launched by NASA in 1992, failed to achieve Martian orbit. The leading candidate as the cause for the failure and loss of the mission is an explosion in the propellant line just as the engine was being prepared to fire in order to capture the spacecraft into orbit around Mars.

Even today, getting a rocket off the pad is not trivial. Within a 12-month period, from the summer of 1998 to the spring of 1999, the United States alone suffered the loss of six rockets. Not all of these losses were caused by the rockets going "boom," but some were. Others were caused by the failure of other critical systems that all must work flawlessly in a very short period of time in order for a launch to be successful.

But what does this have to do with solar sails? The operation of a solar sail is pretty boring when compared to a rocket. To be fair, there are many ways a

Figure 8.1. A Vanguard Rocket failure.

sail may fail. But none of the possible failure modes for a solar sail are as spectacular as a chemical rocket exploding into a cloud of expanding gas. A solar sail contains no combustible material. A failed sail would bear more resemblance to a broken fan than an exploded bomb. This is not just an aesthetic concern. Many people have been killed either onboard or around exploding rockets. The risk of using or being around a rocket is very high. There is virtually no risk associated with using or being near a solar sail.

Figure 8.2. Soviet N-1 rockets would have sent cosmonauts to the Moon.

Another significant issue favoring solar sailing is space debris, which by now surrounds Earth at virtually any altitude up to and around the geostationary orbit (35,786 km high). Most of it is the remnants of either the explosion of launchers' upper stages or spacecraft orbit-maneuver engines, or collisions of satellites with other debris. If a sailcraft fails in orbit about Earth, there is one big piece in area (but not in mass) to be monitored,

Figure 8.3. Ariane 501, the maiden flight of Europe's Ariane 5 launcher, veers off course and is destroyed by its automated destruct system. (Courtesy of ESA) (See also color insert.)

not many millions of particles that eventually spread on a large range of altitude. If ever garbage-collection space vehicles were designed in the future, approaching a failed sailcraft would be much less complicated than locating many small pieces (dangerous indeed) endowed with very high speed (~8–10 km/s) with respect to the chaser system. Apart from that, sail-endowed vehicles orbiting about Earth not only can use the sail during their operational lives, but also can be utilized to enter the upper atmosphere so as to burn at the end of the mission. This would avoid the formation of new big orbital debris (as has been shown by author Vulpetti in a theoretical study for the European Space Agency, June 1997).

Toxic Fumes, Flammable Liquids, and All That Stuff

To further compare a rocket with a solar sail, we will now look at what makes them go: a rocket needs fuel and a solar sails needs light. Fuel is carried onboard the rocket; light for a solar sail is obtained by facing the sun. At first thought, this is a difference, but is it one that matters? Yes, it matters—and it matters a lot.

Rocket propellants range from being relatively benign to caustic and dangerous. Here is a list of some frequently used fuels and the complications that arise from their use:

1. **RP-1**, an early and still widely used rocket fuel, is derived from petroleum. The Saturn V rocket's first stage used RP-1 as it sent men toward the moon. Like its familiar counterparts, kerosene and jet fuel, it is volatile and requires care for safe handling.
2. **Hypergolic** propellants are widely used in deep-space spacecraft because they readily and reliably react with an oxidizer—even without any reaction starter. Given their volatility, which is the very reason they are considered so reliable, they must be handled with extreme care before and during launch and mission operations. Common hypergolic propellants in use today are hydrazine, monomethyl-hydrazine (MMH), and nitrogen tetroxide (N_2O_4). Efficient liquid bipropellant engines for orbital maneuvers utilize MMH and N_2O_4. The U.S. Space Shuttle uses MMH for orbital maneuvering. Hypergolic propellants are also highly toxic. For example, when the debris was being recovered after the Space Shuttle Columbia crashed, volunteers were warned to stay away from any propellant tanks that they might find due to possible contamination from any remaining propellant. In addition, complex and burdensome handling requirements are evoked when spacecraft using hypergolic fuels are being loaded onboard a rocket for launch. The potential for a propellant leak causing worker injury or death is significant.
3. **Cryogenic** propellants are widely used in rockets because of their high efficiency. Unfortunately, they are rather difficult to store because, as their name implies, they must be kept cold. The Space Shuttle's main engines use liquid hydrogen propellant and liquid oxygen oxidizer. Both are gaseous at room temperature and therefore must be kept refrigerated to remain liquids. Hydrogen has a nasty tendency to burn and is notoriously difficult to store due to its low molecular weight and small size. (Those tiny hydrogen molecules easily leak from storage chambers, tanks, and the plumbing associated with a rocket engine).
4. **Solid** propellants are typically less volatile than their liquid counterparts, but they have their own problems. For one, they cannot be easily throttled or stopped. Once a solid rocket motor ignites, it will most likely continue to burn until the fuel runs out. This kind of performance might be good for an earth-to-orbit rocket, but is of almost no use for deep-space maneuvering where precise and flexible rocket performance is required.

A solar sail uses no fuel; therefore, there is zero risk from toxic fumes or fire. There are no cryogenic components, no propellant tanks, and relatively few safe-handling concerns during prelaunch operations and integration of the spacecraft with the launch vehicle.

Complicated Plumbing, Big Tanks, and Turbo-Machinery

Rocket propulsion is complex. It is not called "rocket science" for nothing! A typical liquid fuel rocket engine has many moving parts, pumps, valves, lines, and chambers, every one of which must work perfectly or the rocket may become the bomb mentioned at the beginning of this chapter.

While not used for propulsion in deep space, the Space Shuttle main engine (SSME) is nonetheless a good example (perhaps the best example) of both rocket performance and engineering complexity. A picture of the Boeing SSME is shown in Figure 8.4. According to NASA, the SSME "operates at greater temperature extremes than any mechanical system in common use today." The temperature of liquid hydrogen fuel is 20°K (or −423°F), the second coldest liquid on Earth. When the hydrogen is burned with liquid oxygen, the temperature in the engine's combustion chamber reaches almost 3600°K (or +6000°F)—that's higher than the boiling point of iron. Finally, "the SSME high-pressure fuel turbopump main shaft rotates at 37,000 revolutions per minute (rpm) compared to about 3000 rpm for an automobile operating at 27 m/s (or 60 miles per hour)." Rocket engines used for in-space applications do not have to have this level of performance, but common design issues remain:

- Liquid fuel must be stored, pumped, and mixed with oxidizer in a controlled fashion.
- Pumps (rotating machinery) must operate at high speeds, sometimes after being dormant in space for years.
- Combustion chambers must be capable of sustaining long-term operations at extremely high temperatures.

In 2004, the rocket engine used by the Cassini spacecraft to enter into Saturn's orbit had to fire for more than 90 minutes after being mostly dormant since its launch 7 years previously. The engine performed as designed, but as Project Manager Bob Mitchell is quoted as saying before the engine was ignited, "We're about to go through our second hair-graying event," meaning that the successful firing of the onboard chemical rocket engine and the uncertainty as to whether it would work as designed was a

Figure 8.4. A test of the Space Shuttle's main engines. (Courtesy of NASA) (See also color insert.)

very stressful event for the management and technical teams. Todd Barber, Cassini's leader for the propulsion system, called that system, "a plumber's nightmare." So complicated was the engine that a complete backup was launched onboard in case the primary were to fail. The cost of carrying a full backup engine is far more than the money spent to build it. The mass required for the spare engine might have been used to accommodate more science instruments.

Chemical propulsion systems are also big. In addition to the rocket itself, fuel tanks and propellant feed systems are needed. Taken together, the overall propulsion system typically accounts for most of the mass of any spacecraft we launch into deep space. This is a significant point—most of the spacecraft mass is devoted to propulsion, not to science. Anything that can reduce the mass of the propulsion system will allow more science to be included.

Solar sails comprise but a small fraction of a spacecraft's overall mass. They are relatively lightweight and require no big tanks, no fuel lines, and no complex, rapidly rotating turbomachinery.

Running Out of Gas (Not!)

Another important difference between a chemical rocket and a solar sail is their inherent mode of operation. A chemical rocket typically operates for a very short period of time and delivers a large amount of thrust. A solar sail will operate for years and provide only a small, but continuous, thrust.

Most rocket engines are designed to burn once and for a very short period of time (typically tens of minutes or less). They are able to work this way because their very high thrust imparts a large change in velocity to the spacecraft, rapidly accelerating it. The penalty for high thrust is efficiency. A rocket engine operating at peak performance is limited in what it can do by the chemistry of the reaction that drives it. Most chemical reactions that occur when a propellant is burned in the presence of an oxidizer are inherently inefficient. Rockets provide a big boost, and then fizzle as they run out of gas. It is simply impossible to carry enough fuel to overcome this inherent inefficiency. The more fuel you carry, the more fuel you require to simply move (accelerate) the mass of the additional propellant. This is a no-win scenario that limits what missions can be accomplished by rocket propulsion.

A solar sail produces a very small thrust, but it can do so continuously as long as it is near the sun. Rather than burn for a few minutes and then coast for years toward some interplanetary destination, a solar sail–propelled craft will slowly accelerate and then continue to do so as long as there is sunlight. In other words, while the chemical rocket will put a spacecraft on a mission trajectory quickly, a solar-sail craft will slowly catch up and then overtake its chemical counterpart, resulting in a much higher interplanetary velocity than can be achieved chemically. In addition, the ratio between the thrust acceleration and the solar local gravitational acceleration keeps almost constant independently of the Sun-sailcraft distance. Of course, the closer the sail is to the Sun, the higher its speed gain. Combining both features represents a significant advantage from the trajectory control viewpoint.

However, solar sails do not have the thrust required to lift a payload from the surface of Earth into space, nor could they operate in Earth's atmosphere due to their high atmospheric friction. In this region, rocket propulsion clearly wins. Once the craft is out of the atmosphere and into deep space, however, the calculus changes and the long-life, low-weight aspects of a solar sail propulsion system clearly give it the advantage over its chemical propulsion brethren.

The general lesson that one can learn from the above comparison is that future ambitious space missions could be carried out in the multiple propulsion mode. Each propulsive mode could be appropriately selected, *and fully optimized*, according to the ambient where it shall operate, from the launch from ground to the final destination through a number of intermediate phases. So far, it may sound strange. Such a simple, but powerful, concept about spaceflight has not yet been realized in practice (apart from few exceptions regarding mission classes with targets close to Earth).

Exploring and Developing Space by Sailcraft

9

Sailcraft offer unique opportunities to space-mission planners. Some of these possibilities will be exploited in the near future, others within a few decades, and some in the more distant future. We consider near-term mission possibilities first.

Near-Term (2010–2020) Sailcraft Mission Options

The earliest operational solar-sail missions will demonstrate the utility of this space-propulsion technique. Most likely, these missions will be directed toward destinations within a few million kilometers of Earth.

Solar Storm Monitoring

Most are not aware that the 'good old stable Sun' is not so stable. Several times a year, and more frequently during the peak of its 11-year cycle, the Sun emits Earth-sized bursts of high-energy plasma (a mixture of charged particles, typically protons, electrons, and alpha particles) into space. These enormous plasma bursts, often referred to as "solar proton events" by physicists and "space storms" by the media, speed outward from the Sun at 500 to 1000 km/s. Eventually, some cross the orbit of Earth and wreak havoc with Earth-orbiting spacecraft. We don't directly notice their impact, unless we happen to be living near the North or South Pole, in which case we would see an increase in auroral activity as some of the radiation penetrates the ionosphere and gets trapped along the planet's north/south magnetic field lines. The resulting auroral glow, often called the "northern lights," are often spectacular as the ions spiral along the magnetic field lines deep into the atmosphere, ionizing atmospheric oxygen and causing it to glow a brilliant green. A brilliant light show, but not a threat to human life and activity—correct? Well, this would have been correct if stated 100 years ago, but not today. It is not correct in our age of dependence on electricity and on spacecraft for communication, weather forecasting, and national defense.

G. Vulpetti et al., *Solar Sails*, DOI: 10.1007/978-0-387-68500-7_9,

Spacecraft using solar sails will help us mitigate the risks posed by such storms.

When a space storm reaches Earth, interesting things happen to the protective bubble around our planet called the magnetosphere. To understand this, and how it is germane to our civilization, requires first some understanding of the magnetosphere itself. Recall that all magnets have both a north and a south pole, between which is generated a magnetic field. You cannot directly see this field, but you can observe its effects. For example, if you have two bar magnets and attempt to touch their two north poles together, you feel a repulsive force. Conversely, if you take the north pole of one magnetic and attempt to gently touch it to the south pole of another you, have the opposite problem. They will attract each other, and you must exert force to keep them from coming together too quickly. Any magnet generates a magnetic field; it is through this field that adjacent magnets interact; they interact by attracting or repelling each other, depending on their orientation.

Earth itself generates the equivalent of a bar magnet somewhere in its interior, with the magnetic north pole being near the "top" (or north spin axis point on the planet) and the south pole on the "bottom." (In actuality, Earth's spin axis "north pole" and its magnetic north pole do not physically coincide. They are offset by approximately 11 degrees.)

The next piece of the puzzle that must be understood in order to explain the interaction of a space storm with Earth is the ionosphere. The ionosphere is a region of the atmosphere that begins at an altitude of roughly 80 kilometers from the surface. The atmospheric density has decreased at this altitude to the point where sunlight strips electrons from their parent atoms (typically oxygen) and they exist for extended periods of time as "free electrons" before they collide and recombine with some other atom. The flows of ionized oxygen (and other) atoms and these free electrons form plasma.

Another interesting property of charged particles is that they are affected by both electric and magnetic fields. A charged particle, like an electron, in the presence of a magnetic field will experience a force that causes it to move in a direction perpendicular to both its initial direction of motion and to the magnetic field lines. The magnetic field exerts a force on the ions and electrons in the plasma that results in them spiraling along Earth's magnetic field lines, bouncing back and forth between the North Pole and the South Pole.

Earth's magnetic field, second only to Jupiter in strength within the planets of the solar system, acts as a shield against these intense solar storms, which repeatedly diffuse in the solar system. Without it, life on Earth might not be possible—and it certainly would not be what we see today.

High levels of radiation can certainly harm living things, but it also damages or disrupts the function of electronic systems. Complex systems, such as those found in spacecraft, are especially vulnerable. Satellites launched into space are designed to minimize the effects of these storms and by-and-large, do so successfully. The easiest way to protect against the harmful effects of the ions in the solar storm is by adding shielding mass. Mass, simply put, blocks the ions from reaching whatever is behind it. Unfortunately, with launch costs near $15,000 per kg ($7000 per pound), most satellite users don't want to spend a lot of money adding mass to whatever it is they are launching into space. They instead prefer to either save the money or use whatever extra mass they have available to add more payload (whether it be transponders or science instruments), thus increasing either their revenue or overall science return.

Unless the owners and operators of Earth-orbiting spacecraft do something to mitigate the effects of these storms, damage will occur. The loss of a satellite might seem at first to be an esoteric risk that affects "someone else." Instead, imagine the loss of weather satellite coverage for an extended period of time, including the hurricane season; the ability to accurately predict the location of landfall for a category 4 or 5 hurricane declines to the point that major population centers must be evacuated just because we don't precisely know the track of any particular storm and people may be in its path.

Companies and whole industries use the global positioning system and other satellite assets to accurately manage their inventories and track shipments. Corporate managers plan their business strategy and make decisions based on where certain products or materials are located at any given time. With a sudden loss in this capability, millions or even billons of dollars might be jeopardized.

Cable television, now estimated to be in 68 percent of television-equipped U.S. households, is also at risk. After all, the cable only carries the television signal from your local cable company to your living room. The cable company gets the television signals from satellites located about 36,000 kilometers above the equator. If the satellites go out, the cable companies go out of business.

Perhaps most importantly, the loss of our military and spy satellites would leave whole countries vulnerable to a surprise attack. Knowing that the spy satellite infrastructure is "down" might be a very tempting opportunity for an adversary to take advantage of.

There are other, more down-to-earth impacts spawned by these storms, especially for those living at northern latitudes. Recall that charged particles moving through a magnetic field will experience force acting upon them. So also will a moving magnetic field induce an electric current in a wire. Electrical utility wires (particularly those hanging from telephone poles at northern latitudes) will feel the effect of the solar storm as Earth's magnetic field is compressed, varying in intensity with time. This changing magnetic field induces current flow in the wires, creating spurious currents that knock out transformers and otherwise disrupt or shut down the transmission of electrical power. This is a real effect and it has happened.

Fortunately, in addition to adding lots of mass to the spacecraft, there are two other ways to mitigate the effects of these storms, and both require some sort of advanced warning of an impending storm. One is to turn off particularly vulnerable spacecraft systems when the worst of the storm arrives, and power back up after the storm is over. The other would be to reorient the spacecraft so as to maximize any onboard spacecraft mass between the most vulnerable systems and the ionizing radiation in the plasma for the duration of the storm.

To provide at least some warning of impending solar storms, the U.S. National Oceanic and Atmospheric Administration (NOAA) and NASA placed the Advanced Composition Explorer (ACE) spacecraft at one of the so-called Earth/Sun Lagrange points. In the 17th century, Italian mathematician Giuseppe-Luigi Lagrange (who worked in France for 27 years) discovered that there exist regions in space where the gravitational attraction of the Sun and Earth mostly cancel each other out, meaning that a spacecraft placed in one of these regions will likely remain there unless acted upon by some outside force. These regions (around the so-called *libration* or L-points) are not 100 percent gravity or disturbance free, so some spacecraft propulsion is required to remain within them (Fig. 9.1). The fuel required there, however, is much less than would be required should these regions not exist. The ACE spacecraft is located at the Earth/Sun L1 point (1.5 million kilometers from Earth) and monitors the Sun for solar-proton events (Fig. 9.2). It detects such an event when the light from a solar flare associated with the event strikes its detectors; the spacecraft then sends a radio signal back to Earth. The radio transmission signaling the impending storm reaches Earth about 1 hour before the ionizing radiation because light travels faster in the vacuum of space than do the charged particles in the storm. Unfortunately, 1 hour is not much time, in general, even considering the very high number of aircraft in flight and the complexities of the various human activities that could be seriously degraded or halted by solar storms.

Here's where the sail comes in. The next generation of L1-based solar observatories will use conventional rockets to escape Earth's influence and decelerate at L1. Then, the craft will deploy a modest solar sail, perhaps 50 to 100 meters in diameter. Applied to correct for the various gravitational

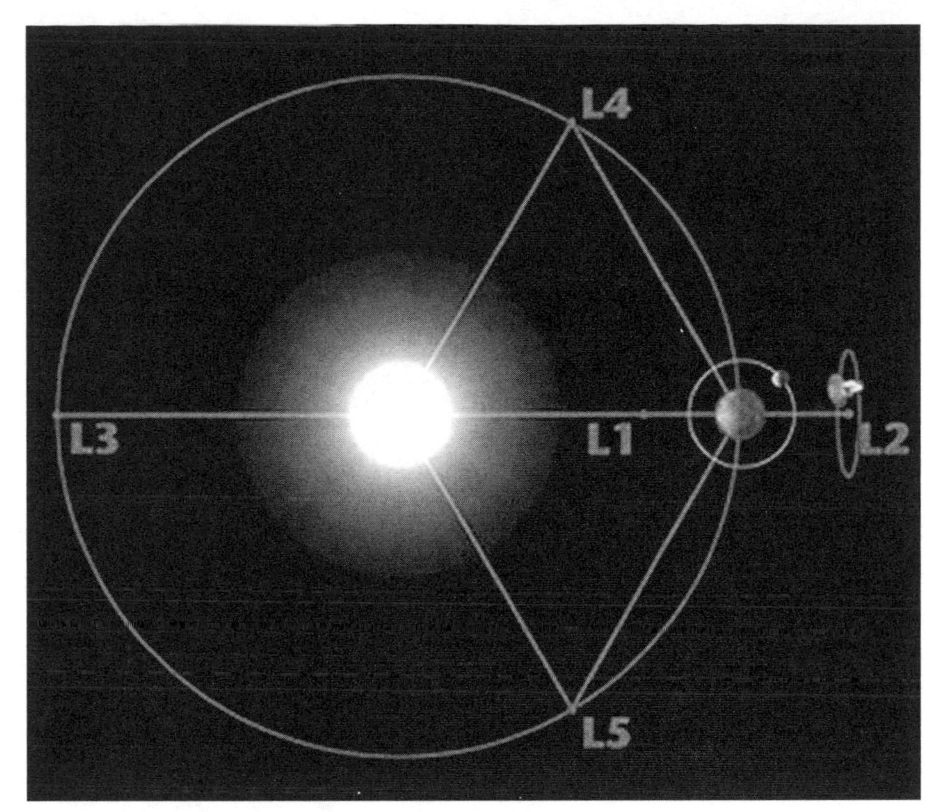

Figure 9.1. The Lagrange points of the Sun-Earth system.

influences working to drag the solar monitor off station, the sail could maintain spacecraft position with a great reduction in on-board fuel requirements, which translates into longer spacecraft operational lifetime and lower costs.

Even better, a sailcraft can be positioned in a direct line between the Sun and Earth, remaining in this otherwise propulsive-intense location, and can be available to provide earlier warning of impending storms. The ship must thrust continuously to remain on station—a task ideal for a solar sail.

NASA is considering a mission using a solar sail to either replace the ACE spacecraft or as a complement to its replacement. The potential mission has been called many names, from "Geostorm" to its current incarnation, "Heliostorm." Trade studies to determine the optimal science instrument complement, spacecraft, and solar-sail propulsion-system characteristics are ongoing, and will likely continue until the mission is approved for flight. An early Heliostorm concept would use a square sail, 70 meters on each edge, with a total mass of <200 kilograms to accomplish the mission goals. It would be launched from Earth in a relatively small rocket, such as a Pegasus

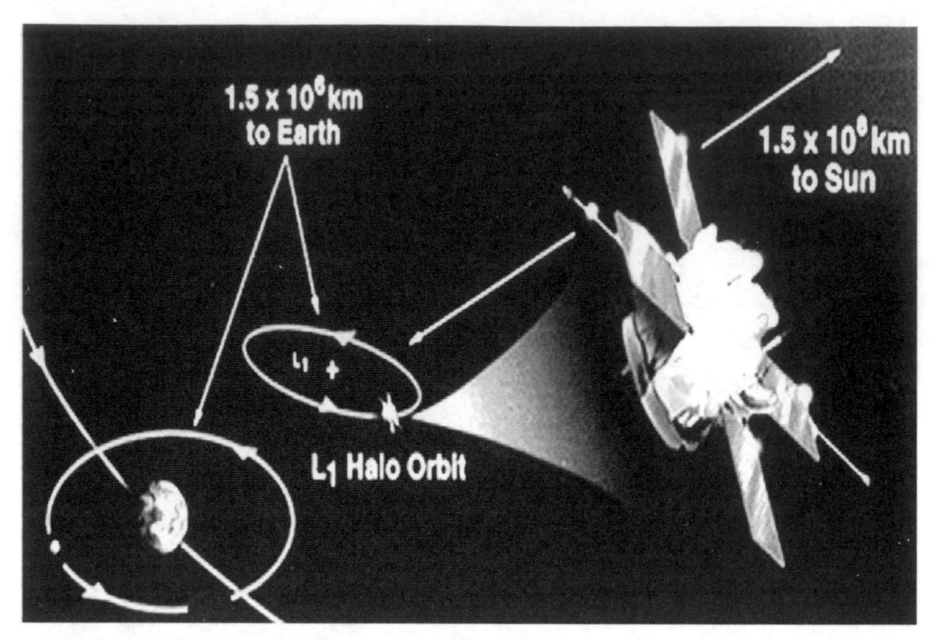

Figure 9.2. The ACE spacecraft around L1.

or Falcon, and propelled to 0.98 astronomical units (AU) by a conventional chemical rocket. The sail would then deploy and operations commence.

An advanced version of the Heliostorm concept (based on advanced technology) would consist of a sailcraft with a total mass of 300 kilograms and a circular sail 230 meters in radius. Such a spacecraft could orbit the Sun stably at 0.70 AU (never being captured by Venus) in the ecliptic plane with a period of exactly 1 year. Once put 0.30 AU from the Earth sunward, its mean position with respect to Earth would not change, and the space-storm warning time would range from 16 to 31 hours. (A technical explanation of this mission concept can be found in Chapter 17.) Such time would be enough even for astronauts who will be working on the Moon, far from their lunar base.

Pole Sitters

Another space mission of interest that could be implemented in the near term is the terrestrial pole sitter. Using the thrust on a sail provided by sunlight to balance Earth's gravitational attraction, a pole sitter is just that—a sail-propelled spacecraft that appears to "sit" above one of the Earth's poles. This is a high-latitude analogue to the geosynchronous position of most communication satellites, which are permanently stationed 35,786 kilometers above the equator. The period of such an orbit is exactly one Earth rotation period (about 86,164 s) and the satellite will apparently

The way that part of the solar wind reaches Earth's upper atmosphere is somewhat complicated and not yet completely explained. As discussed in Chapter 7, a significant step in our understanding of the solar-wind interaction with Earth's magnetosphere has occurred in recent years based on key observations from a number of modern satellites designed for such an aim—the missions IMAGE (NASA) and CLUSTER (ESA). For the first time, there has been the observational evidence of something conjectured some decades ago: the magnetic reconnection. The solar wind carries the lines of the interplanetary magnetic field (IMF); when the IMF direction is opposite to that of the field of Earth's magnetopause (the magnetosphere boundary that acts as our magnetic shield), the lines of the two magnetic fields can first break and then reconnect with each other in such a way that one or more enormous "cracks" (typically larger than Earth) are produced in the magnetopause. The solar wind now slides along the terrestrial magnetic lines down to the ionosphere. In going down, the plasma tube area decreases to sizes typically equivalent to Japan. The unexpected feature of such magnetopause cracks is that they can stay open for many hours, thus provoking *severe space storms*.

remain stationary above the same location on Earth's surface and serve as a convenient target for radio beams.

A terrestrial pole sitter would be situated in the sky as near to the pole star's location as possible. Rather than rotating at the same angular velocity as Earth's surface, it would have a relatively constant location on the celestial sphere. Thus, designers of telecommunication, Earth-resource, navigation, and weather satellites designed to serve high-latitude users cannot use this convenient orbit since high-latitude ground stations find geosynchronous satellites to be below or near their horizon.

Although high-latitude pole sitters are certainly possible, they will have certain annoying consequences for telecommunication customers. Studies indicate that solar-sail pole sitters will function best if situated at the Moon's distance or beyond. This greater distance would introduce a longer time delay in telephone calls via pole-sitter spacecraft. This will not please all customers!

The ability to hover over a single area of Earth would be highly desirable for those monitoring the environment. Instruments placed on such craft would be able to continuously monitor local weather and environmental conditions. Military users would also benefit from the ability to continuously observe the activities of a potential adversary.

Magnetospheric Constellations

An additional near-term, near-Earth possibility is to launch a number of mini–solar sails (not the micro-sails introduced in Chapter 7, but decidedly larger), or "solar kites" aboard the same rocket. Equipped only with miniaturized communications and navigation gear and instruments to monitor space radiation and fields, these craft could use solar sails to cruise through Earth's magnetosphere, between, say, 2000 and 50,000 kilometers above Earth's surface. This scientific "constellation" would yield real-time and synoptic data about variations in Earth's magnetic field and radiation belts.

Target-Variable Magnetospheric Missions

With regard to deeper scientific exploration of Earth's magnetosphere and, at the same time, to experiment with solar-sail technology and study its problems, the European Space Agency is considering a mission named Geosail. In reality, the primary goal of Geosail should be the full demonstration of solar sailing, though the sail area should amount to about 1900 square meters (m^2) (2.5 times the area of the baseball diamond). Full demonstration means mainly sail packing, in-orbit sail deployment, sail attitude acquisition and control, sail-state monitoring, using solar-pressure acceleration for continuous orbit change, sail attitude maneuvers, sail detachment, and even (indirectly) observing sail's materials degradation. The mean orbit of Geosail would be well beyond the geostationary orbit, between 70,000 and 150,000 kilometers; its perigee should touch or cross the near magnetopause, whereas its apogee should dive in the magneto-tail. Such orbit shall be high *variable*, not only because of the gravitational perturbations caused by the Moon and the Sun, but mainly because the magnetopause continuously changes also in orientation. As a point of fact, the solar wind moves radially from the Sun, and Earth revolves about the Sun; thus, the magnetosphere's elongated shape axis varies to be always aligned with the Sun. The scientific goal of Geosail is very important; it could be considered an appropriate continuation of the Cluster mission. Geosail should last 3 to 5 years with a mass lower than 200 kilograms, thanks to the solar-sail propulsion.

Solar Polar Imager

Solar sails are especially effective at performing missions that otherwise require a large amount of propellant. A particularly propulsive-intense maneuver is required to change the orbital inclination of a spacecraft, whether it orbits the earth, another planet, or the sun. The orbit's inclination is simply a measure of the angle the orbit plane makes with respect to some reference plane, which usually is either Earth's equator or the ecliptic. (Note that the ecliptic is a plane, not an orbit, although the terms *earth orbit* and *ecliptic* often are used interchangeably.) Moving from the initial launch orbit to another "angle" in orbit is very difficult, and conventional propulsion systems are limited in performing this maneuver by the amount of fuel they can carry.

Taking advantage of a solar sail's virtually unlimited ability to provide thrust, scientists are eager to place a spacecraft into a highly inclined orbit around the Sun in order to study what happens near its poles. Current observations of the Sun are limited to spacecraft launched from Earth, which remain nearly in the ecliptic (because they are launched from Earth, which inhabits the ecliptic), limiting our views to those near the solar equator or its mid-latitude regions. The proposed mission to study the Sun's poles is called, the Solar Polar Imager, and it can only be realistically implemented with a solar sail propulsion system. The Solar Polar Imager spacecraft would benefit from the $4\times$ increase in solar propulsive thrust available from operating at 0.5 AU. While the proposed mission is just a concept at this time, studies show that current solar-sail technology could be used to implement the mission with a square, 3-axis stabilized sail no more than 150 meters on a side.

L-1 Diamond

Taking the Heliostorm concept a step further and increasing the number of sunward, solar-observing solar-sail–propelled spacecraft in orbit around the sun would dramatically increase our understanding of our star. The L-1 Diamond mission is one proposal to achieve simultaneous, multiangle solar observations providing all the advantages inherent in having multiple views of complex phenomena. L-1 Diamond is proposed to be a constellation of four spacecraft working together to gather information about the sun and the solar environment.

Three of the spacecraft would fly in triangular formation around the sun. The fourth spacecraft would be located above the ecliptic, looking downward. Again, this mission could be achieved with first-generation solar sails.

Mid-Term (2020–2040) Sailcraft Mission Options

Moving forward a few decades, we can reasonably expect major improvements in sail technology. Various sail structures and unfurlment techniques will have been perfected. Sails will be thinner, stronger, and more temperature resistant. A number of exciting mission opportunities could be implemented during this time frame. One of these is the possibility of formation flying with a comet and returning comet samples to Earth.

Comet Rendezvous

All the major planets and most asteroids circle the Sun in or near the same plane that Earth does—called the ecliptic. The constellations of the Zodiac are arrayed along the ecliptic track on the celestial sphere. Comets, on the other hand, are all distant from the ecliptic. It is very difficult to visit a comet at an arbitrary point of its orbit, because of the very high energy required to shift orbital inclination to match that of the comet. But given months or years, a solar sail in the inner solar system can perform such an inclination-cranking maneuver without the expenditure of an on-board propellant.

It's true that the current, conventionally propelled probes have visited the vicinity of a few comets, but these were short-term flybys (or in some cases fly-throughs) in which the probe traveled past the comet at relative velocities of 50 kilometers per second or more.

A sail-propelled probe could utilize solar radiation pressure to match orbits with a comet and cruise in formation with that celestial object for weeks or months. Samples of comet material could be gathered for later return to Earth.

Particle Acceleration Solar Orbiter

The Particle Acceleration Solar Orbiter would allow close-up imaging (<0.2 AU) and spectroscopic analysis of high-energy solar flares to determine their composition, development, and acceleration mechanisms. Seeing the life cycle of a flare event from close solar orbit will significantly advance our understanding of these events.

Mars Sample Return

Returning a sample from Mars has long been a goal for scientists interested in learning more about the possible development of life beyond Earth. Unfortunately, the complexity and associated high cost with performing this mission seem to push it indefinitely into the future. One aspect of the problem is the fuel required for the return trip to Earth. Getting a spacecraft to Mars requires a large, dedicated launch vehicle. Any sample return mission would have to also include a rocket landed on the surface of Mars to return the sample from the surface back into space. Once back in space, the sample would then have to be transported to Earth. To do this chemically would require multiple rocket launches. We simply cannot launch at one time enough fuel to get our spacecraft into orbit for the Mars ascent rocket, and the propellant required for returning to Earth.

If the mass required for any leg of the trip can be significantly reduced, the cost of the mission would decrease, making it more likely to happen. Solar sails provide a lightweight option for returning the sample from Mars to Earth. The scenario might go something like this: (1) A rocket launches the mission spacecraft from Earth, (2) The spacecraft enters Martian orbit, sending a lander to the surface, (3) The lander collects the sample of interest and sends it back to space using a rocket that accompanied it to the surface, (4) The rocket has a rendezvous with a solar-sail–propelled craft in Martian orbit, transferring the sample, (5) The sailcraft returns the sample to some parking orbit about Earth, (6) An orbital transfer vehicle moves the sample to the future space station. (Alternatively, the sailcraft could return the sample to the lunar base.) In this scenario, the lightweight solar sailcraft replaces the heavy chemical propulsion stage that would otherwise be required to return the sample to Earth for analysis.

Aerocapture Experiments

One method of reducing the cost of some scientific space missions is to utilize a new technique called aerocapture. In performing this maneuver, a space probe must be directed toward a solar system object with an atmosphere, such as the planets Venus, Earth, Mars, Jupiter, Saturn, Uranus, and Neptune, or Saturn's satellite Titan. Instead of using rockets to decelerate for capture from a Sun-centered to planet-centered orbit, the craft grazes the planet's atmosphere. If the orbit is precise, atmospheric friction will decelerate the spacecraft sufficiently for planetary capture.

A number of specialized devices—aeroshells and ballutes (which are a cross between balloons and parachutes)—could be deployed by a space probe performing an aerocapture maneuver. Preliminary studies reveal that certain solar-sail configurations could be applied during aerocapture. An added bonus to sail application in aerocapture, of course, is that the sail can utilize solar-radiation pressure to accelerate a spacecraft, as well as functioning parachute-like to slow one.

One possible sail-aerocapture probe would be a Titan orbiter. After Earth launch, the sail would be utilized to accelerate the spacecraft toward Saturn. Arriving at Saturn, the sail would be utilized as an aerocapture device to steer the craft into a Saturn-centered orbit with an apoapsis (high point) near Titan. The sailcraft would once again apply aerocapture, grazing Titan's atmosphere with a fully unfurled sail, to become a satellite of that tantalizing small world.

Another mission option would involve a sail launch toward Mars, sail-aided aerocapture into Mars orbit, and sail-aided maneuvering in the Mars system. Samples could be returned from the surfaces of Mars's small satellites Deimos and Phobos. If these samples contain ample amounts of water and other volatile materials, later human-occupied ships visiting the Red Planet could utilize these satellites to top off their fuel tanks.

Extrasolar Probes

One might suppose that a low-thrust Sun-pushed gossamer spacecraft will have no application in ventures testing the fringes of galactic space. One would be wrong!

In a maneuver called "sundiving" by science-fiction authors Greg Benford and David Brin, the sailcraft is initially placed in a parabolic or elliptical solar orbit with a perihelion (point of closest solar approach) as close to the Sun as possible (Fig. 9.3).

At perihelion, the sail is pointed toward the Sun, and the craft is ejected from the solar system. Contemporary, Earth-launched sail technology seems capable of achieving solar-system escape velocities as high as 50 kilometers per second (km/s), with or without the giant-planet gravity assists utilized to accelerate Pioneer 10 and 11 and Voyager 1 and 2.

The solar-system escape velocities possible utilizing the sundiver maneuver are far in excess of the velocities of the Pioneers and Voyagers. Within a flight-time of about two decades, a sail-launched extrasolar probe could reach the heliopause—the boundary between solar and galactic influence—at 200 astronomical units (AU) or so from the Sun and measure local field strengths and particle densities. if the probe can survive another few decades—not impossible in light of Pioneer's and Voyager's longevity—data could be returned from the inner focus of the Sun's gravitational lens

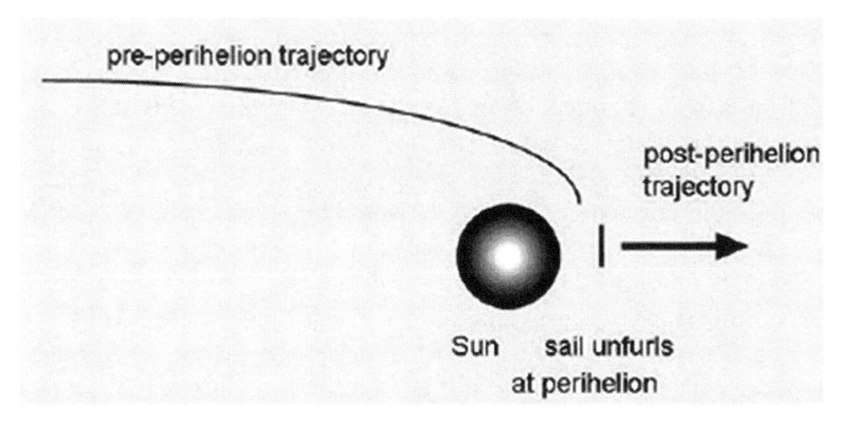

Figure 9.3. A Sundiver Maneuver.

at 550 AU, which would provide a check on Einstein's general relativity theory as a mission bonus.

The possibility to fly by the Sun and get a high escape speed from the solar system is much more than a science-fiction idea. We have a strict mathematical theory, coming from many years of scientific research, which tells us that even without resorting to far future technologies, a sufficiently light sailcraft could be controlled in such a way as to reach the solar gravitational lens with a speed of 120 km/s, at least. Of course, emerging technologies could do excellent things for the designs of whole sailcraft and help us to transform such theoretical results into reality.

Some may conservatively argue that probes to the edges of interstellar space may have little relevance to terrestrial life. But, since mass-extinction events on Earth may be linked to galactic influences, a few such interstellar craft may well be launched as our sail technology improves. And these early flights will only be the start of the sail's flirtation with the galactic abyss. Because the solar sail is scalable, we may view these early efforts as humanity's first true starships.

Far-Term (2040 +) Sailcraft Mission Options

As time elapses, humanity's technological progress is certain to continue. After 2040, a substantial in-space infrastructure may well exist. There may be facilities in near-Earth space where space resources or Earth-launched material can be processed to produce solar sails with near-theoretical-maximum performance.

Larger sails will be possible in this time frame—with dimensions measured in kilometers. And these large, space-manufactured sails will perform better than their Earth-launched predecessors.

Human-Exploration Sailships

Current-technology, micron-thick, Earth-launched sails are not yet up to the support of human exploration of the solar system. These sails are too small to carry the tens of thousands of kilograms necessary to support humans between the planets and exploration gear. Also, sail-implemented missions to Mars (for example) using today's sail technology would be of longer duration than rocket-propelled interplanetary ventures.

But when sail linear dimensions are measured in kilometers and sail thicknesses are in the sub-micron range, all this will change. The sail may then become the most economical means of transport throughout the inner solar system. New constellations of 21st century space clipper ships might be visible in Earth's night skies as they spiral outward toward Mars or the asteroids. The first of these might be rather modest, a mere 800 meters on a side, carrying 5000 to 10,000-kilogram payloads between Earth and Mars on a recurring basis. While too massive to launch from Earth, such a large-diameter sail could be readily made in space to perform this mission without overly stressing the other sail figure of merit—areal density.

Initially, these craft will support exploration missions. But since sailships should be capable of many interplanetary roundtrips without fuel expenditure, human settlements will also benefit from the technology as they begin to grow on celestial bodies beyond the Moon.

Rearranging the Solar System

Although Mars exploration has captured the hearts and minds of the public, altering orbits of some near-Earth objects (NEOs) of asteroidal or cometary origin may be of much greater terrestrial significance.

There are thousands of these bodies scattered between the orbits of Venus and Mars. And it is known that they occasionally whack Earth, with disastrous consequences.

The most famous of these impacts occurred about 65 million years ago in the Yucatan region of Mexico. Eons before the Mayan rulers sported in the warm waters of the Caribbean, the tremendous fires and ash from the impact of this 10-kilometer object may have helped cause the demise of the dinosaurs and the rise of mammals to ascendancy.

But smaller objects, such as the 100-meter-radius object that impacted in Tunguska Siberia in 1908 with the force of a 20-megaton hydrogen bomb, strike much more frequently than NEOs of the dinosaur killer's size—at intervals of few centuries or less.

Although in principle there is a certain (low) risk of Earth-NEO collisions in the course of centuries, do not panic; unlike our ancestors, we can do something about this threat, taking responsibility for our own future.

Although nuclear explosives are certainly an option to divert a NEO targeting Earth, sails present a much more elegant option. A large, thin solar sail deployed at an NEO would increase both the reflectivity and the effective area of the NEO, allowing for a decades-duration alteration in the NEO's orbit, converting a direct hit on Earth into a near miss.

Space Mining

If we are going to explore the NEOs, why not make use of them? Many materials are present in or on these small celestial bodies, including (at least in some NEOs) water.

One way to support an expanding space infrastructure and render it less dependent on Earth is to mine NEOs as we rearrange their orbits and ship the materials back to space-processing facilities near Earth. Solar sails may provide an economical, though slow, method of altering the orbit of an NEO to allow its riches to be mined and exported back to Earth and elsewhere.

The NEO-obtained material may be used to create a geosynchronous ring of solar-power satellites to beam energy back to Earth, rendering the West's current oil addiction obsolete.

Solar-sail freighters, perhaps under robotic control, would make very effective transporters of material from the NEOs. Such an application may prove to be the most economically significant of all sail uses in this time frame.

Oort Cloud Explorers

As a prelude to interstellar travel, space agencies after 2040 may develop an interest in probing the inner fringes of the Sun's Oort comet cloud.

Although some comets occasionally approach the Sun, where they display beautiful comae and tails, most reside in the frigid wasteland beyond the most distant planet of our solar system. Perhaps a hundred billion or a trillion of these ice balls are out there, some as close as a few thousand astronomical units, others as far out as 70,000 or 80,000 AU. Occasionally, a passing star or other celestial influence disrupts some of these objects from their stately orbits and shunts them sunward as a comet swarm.

We have probed some of the comets that regularly visit the inner solar system but it would be nice (and informative) to visit these relics of solar-system formation in their natural realm.

This is a task for the Oort cloud explorer, perhaps the ultimate sailcraft before a true starship. Imagine a sail 100 **nanometers** thick, perhaps a kilometer in radius, which is constructed of material capable of withstanding a perihelion pass of about 0.05 AU (about 10 solar radii). Such a craft could perform a Sun dive and project its payload toward the stars at velocities in excess of 500 kilometers per second.

Although the Oort cloud explorer would take perhaps 2000 years to traverse the 40 trillion kilometer (4.3 light year) gulf between our Sun and its nearest stellar neighbor, it could certainly survey the Oort cloud out to a few thousand AU during its operational lifetime.

The Ultimate Future: Sailships to the Stars!

Interstellar travel—flight to the stars—seems so easy in the typical Hollywood space epic. A ship silently drifts in interplanetary space, and a button is pushed. Marvelously, the local fabric of space-time is warped and distorted. The spacecraft takes an interdimensional shortcut across the universe, emerging instantly into normal space near a star many trillions of kilometers distant from our solar system!

If only it were so easy in the real world! Such interdimensional shortcuts are possible in theory, but not easily achieved in practice. To warp space effectively, we might need the mass of a star squeezed into the volume of a small terrestrial city—a so-called black hole. Yes, black holes may be shortcuts to distant realms of space and time, but tidal effects would doom a spacecraft foolish enough to approach one closely.

We might consider using angular momentum (spin) or magnetic fields to replace such a gravitational singularity, but how do you keep a structure from blowing apart if it must be spun at half the speed of light to produce an angular-momentum-induced space warp. And to do it magnetically might require production of impossibly strong magnetic fields.

If only it were as easy to take an interdimensional shortcut as portrayed by Hollywood! Many physicists have calculated that contemporary physics actually forbids such techniques, which are based on the assumed existence of exotic matter having *negative* energy density (not to be confused with antimatter). In addition, even if we could produce a tunnel through space—a wormhole—there are stability issues. The energy of the known universe might be required to stabilize the thing long enough for a ship to pass through! Recently, some physicist computed a much lower amount of stabilization energy, but still incredibly high for what we can manage.

But interstellar travel is still possible, even if space warps are quite unlikely. Real starships will be slower than celluloid craft, and travel times will be longer. Before considering application of the solar sail to interstellar travel, let's briefly examine some of the other approaches that have been suggested.

Relativistic Starflight

All right, so instantaneous interstellar travel seems to be beyond us. But what about flight at relativistic or near-optic velocities, close to 300,000 kilometers per second. Even though travel at near light speed would take years or decades from the point of view of Earth-bound observers, even to near stars, special relativity predicts that such flights will be much shorter from the point of view of on-board crew members.

When I.S. Shlovskii and Carl Sagan published their classic, *Intelligent Life in the Universe*, in the 1960s, they noted that only two modes of relativistic travel seemed physically possible. These are the antimatter rocket and the hydrogen-fusing ramjet. Although their operation would not violate the laws of physics, there are serious technological and economic limitations to the near-term development of these travel modes.

Every elementary subatomic particle has a corresponding antiparticle (see Chapter 3). Put some matter and a corresponding mass of antimatter together and—boom! All the matter and antimatter is instantly (and explosively) converted into energy. The matter-to-energy conversion efficiency of the matter–antimatter reaction is more than 100× greater than the best we can do with nuclear fusion and fission.

So all we have to do, conceptually, is load our interstellar rocket with lots of hydrogen and an equal mass of antihydrogen. If the matter and antimatter are allowed to slowly interact, the reaction can accelerate the craft to relativistic velocities.

But there are two big problems. First is the economics of antimatter production. Yes, we can produce tiny quantities—nanograms per year—of the stuff in our most energetic nuclear accelerators. But the cost is staggering. If the entire U.S. economy were devoted to the production of the stuff, even allowing for economies of scale, it is doubtful that we could produce a gram in a decade.

Even if a breakthrough alleviates the economic issue, there is another problem. How do we safely store the stuff for years or decades during the starship's acceleration process? Remember that if even 1 milligram of antimatter comes in contact with the storage chamber (which is constructed of normal matter), the ship will instantly self-destruct!

In principle at least, the hydrogen-fusing ramjet is a more elegant solution. There are plenty of ionized hydrogen particles—protons—adrift in the interstellar medium. A properly configured electromagnetic field (a so-called "ramscoop") could conceivably be utilized to collect these over a thousand-kilometer radius from the interstellar medium in front of a starship. These collected particles could then be directed into an advanced nuclear-fusion reactor and joined together (fused) to create helium and energy. The reaction energy could be applied to the helium exhaust to accelerate the starship up to relativistic velocities.

But as with the antimatter rocket, there are two major issues to constructing a ramjet. In this case, both are technological. First and foremost is the low reactivity of the proton–proton reaction. While it is true that almost all stars, including our Sun, radiate energy produced by proton fusion, this reaction is many orders of magnitude more difficult to achieve in the laboratory than thermonuclear fusion reactions used in the hydrogen bomb and our experimental fusion reactors. Barring a major breakthrough, we may never be able to tame proton fusion without carrying around a stellar mass—a somewhat inelegant approach to interstellar travel.

Even though other reactions could be used to propel slower ramjet derivatives, there is a secondary technological issue. Most electromagnetic ramscoop designs are much better at reflecting interstellar protons than collecting them. It is far easier to design an electromagnetic drag sail to slow a speeding starship than a ramscoop to collect fuel from the interstellar medium.

So we will abandon relativistic starflight concepts from our consideration. What a pity—but we still could have "slow boats" that would take centuries to cross the gulf between our solar system and its nearest stellar neighbors.

The Nuclear Option

The first feasible method of interstellar travel to emerge is a daughter of the Cold War. First as a space interceptor and then as a backup to the Saturn V Moon rocket, the U.S. Department of Defense and NASA considered Project Orion—a spacecraft propelled through space by the thrust of exploding nuclear devices. Tested with conventional explosives (since atmospheric nuclear detonations are prohibited by international treaty), a subscale Orion prototype was successfully flown during the 1960s and is on permanent display in the Smithsonian Air and Space Museum in Washington, DC.

With the technique of nuclear-pulse propulsion demonstrated, physicist Freeman Dyson moved the concept to its theoretical and economic limits in an epochal paper published in 1968 in *Physics Today.* If the Cold War thermonuclear arsenals of the U.S. and the Soviet Union had been devoted to the propulsion of huge Orions constructed in space, small human populations could be transferred to neighboring stellar systems. Travel times to the Sun's nearest stellar neighbors—the Alpha/Proxima Centauri system—would be in the range 130 to 1300 years.

Of course, no nuclear power can realistically be expected to unilaterally donate its arsenal to the cause of human advancement. So the British Interplanetary Society commenced Project Daedalus in the early 1970s to evaluate the possibility of a sanitized version of nuclear-pulse interstellar propulsion.

The Daedalus motor would employ the concept of inertial fusion, a technique that is currently approaching laboratory realization. Small pellets of fusible isotopes, preferably deuterium and helium-3, would be ejected into the craft's combustion chamber, at the focus of laser or electron beams. These beams would compress the pellets and raise their temperature to the point at which thermonuclear fusion can occur. One-way interstellar travel time for small human communities would be measured in centuries, and robot probes would be faster.

But there was one catch. Helium-3, although abundant in the Sun, is extremely rare on Earth. To implement Daedalus, we would have to develop a space infrastructure capable of locating and mining this isotope from resources such as giant planet atmospheres, the solar wind, or possibly the lunar regolith.

If we wish to conduct early interstellar ventures, Daedalus is not practical. But, surprisingly, the solar sail provides an alternative propulsion possibility.

Solar-Sail Starships

You might think at first that the solar sail is useless in the dark void of interstellar space. After all, today's sails are flimsy affairs capable of small accelerations—typically one-ten-thousandth of an Earth surface gravity (0.0001 g).

But recall this—solar flux is an inverse-square phenomenon, meaning that as we halve the distance between the sail and the Sun, the sail's acceleration increases by a factor of 4. If we can unfurl our sail very close to the Sun, then accelerations of 1 g or higher are possible (but only there).

Before 1980, two American research teams were independently evaluating the feasibility of solar-sail starships. Some of the research was performed as part of a NASA Jet Propulsion Laboratory (JPL) study—the TAU (thousand astronomical units) probe, an interstellar precursor probe, departing the solar system at 50 to 100 kilometers per second. Too slow to reach the nearest stars in less than about 13,000 years, TAU would sample particles and fields in the nearby interstellar medium and perform astronomical observations.

Although the favored propulsion system for TAU was the nuclear-electric drive in which a fission reactor's energy is used to ionize and accelerate argon atoms, a solar sail unfurled near the sun was considered as a backup mode of propulsion. Unfortunately, the senior analyst on this aspect of the study, Chauncey Uphoff, was permitted to publish his star-sail extrasolar probe results only as an internal NASA memo.

At about the same time, author Gregory Matloff, in collaboration with Michael Mautner and Eugene Mallove, was independently evaluating solar-sail starship propulsion as an alternative to nuclear pulse. Most of this work was published during the 1980s as a series of papers in the *Journal of the British Interplanetary Society (JBIS)*

An optimized interstellar solar sail probably would be constructed in space using a nanometers-thin monolayer of a highly reflective, temperature-tolerant material—possibly a metal such as beryllium, aluminum, or niobium. The sail would be affixed to the payload utilizing cables with the tensile strength of diamond or silicon carbide.

In operation, a partially unfurled sail might be mounted behind a chunk of asteroid that has been machined to serve as a sunshade. The sail and occulting sunshade would then be injected into a parabolic solar orbit with a perihelion solar distance measured in millions of kilometers.

Approaching perihelion, the partially unfurled sail would emerge from behind its sunshade and be rapidly blown from the solar system. As the solar distance increases, the sail could be gradually unfurled and ballast released to control acceleration.

Analysis revealed that acceleration times measured in hours or days were possible. By the time the ship reaches the orbit of Jupiter, the sail could be furled, since acceleration has fallen to a negligible value. The sail could be used as cosmic ray shielding and later unfurled for deceleration. Flight times to Alpha Centauri, even for massive payloads that could carry human crews, could approximate a millennium. Of course the hyperthin sail sheets required to "tow" such large, multimillion-kilogram payloads, would be enormous—in the vicinity of 100 kilometers.

One way to increase performance of a sail-equipped starship is to "park" a solar-pumped laser or microwave power station within the inner solar system and use this device to beam collimated energy to a sail-equipped starship very far from the Sun. This approach is considered in more detail in the next chapter.

Further Reading

P. Gilster, *Centauri Dreams*, Copernicus Books, 2004.

E. Mallove and G. Matloff, *The Starflight Handbook*, Wiley, New York, 1989.

G. L. Matloff, *Deep-Space Probes*, 2nd ed., Springer-Praxis, Chichester, UK, 2005.

G. Vulpetti, *Overview of advanced space propulsion via solar photon sailing*, lectures presented at the Aerospace Engineering School of Rome University, Italy, May 2005, downloadable from http://www.giovanni vulpetti.it.

10 Riding a Beam of Light

The single most important characteristic of a solar sail is its power source—the Sun. The Sun supplies a continuous source of sunlight, providing the gentle push that makes a solar sail such a useful propulsion system. Unfortunately, the Sun is also the limiting factor in the overall usefulness of a solar sail. When a spacecraft gets far from the sun, there is simply not enough light available to provide additional propulsion. Recall the "inverse square law" discussed previously. In deep space, the Sun is essentially a point source, with sunlight radiating away from it in all directions forming an ever-expanding sphere of light. Since the total amount of light from the Sun is the same when the expanding light sphere reaches the orbit of Mercury, Venus, or Earth, we are not "losing" sunlight. What we are doing, however, is reducing its intensity. The amount of sunlight may be the same, but the surface area of the sphere is much larger the farther you get from the Sun. The only way that the amount of sunlight can remain constant, which we intuitively know it must, yet cover a much larger area, is for the amount of sunlight per unit area to decrease. And decrease it does; as the distance from the Sun doubles, the amount of sunlight falling on a 1-square-meter area on that sphere drops to one-fourth of its previous value. The distance is doubled, and the amount of light is reduced by a factor of 4. Since 4 is 2 squared, this predictable decline in sunlight is governed by the inverse square law and holds true no matter how far away from the Sun the sphere of light travels. If you measure the total amount of light falling on a 1-square-meter area of sail and then quadruple the distance, the amount of sunlight falling on that same sail drops to 1/16 of its previous value: $4^2 = 16$. As we move away from the Sun, the push our sailcraft receives drops rapidly.

Thanks to Newton, we understand that a sail craft won't slow down when the sunlight dims. It will continue moving with whatever velocity it achieved during its acceleration phase until some outside force acts upon it. For a sailcraft targeted to deep space, this might mean that the sail continues on its journey for thousands or millions of years. Without light, it will not continue to accelerate and move with an ever-increasing velocity. If we want to use a

G. Vulpetti et al., *Solar Sails*, DOI: 10.1007/978-0-387-68500-7_10,

sail to reach the stars in a reasonable amount of time (from a human perspective), this simply will not do. Using sunlight alone, with the largest, thinnest sail we can imagine, and with a very close solar approach, a sailcraft will take at least one thousand years to reach the nearest star.

Clearly, we must find a way to change the rules of the game and make sure our sailcraft has an ever-constant beam of sunlight available so that it can continue to accelerate to higher and higher velocities—making journeys into interstellar space possible with a trip time of less than a thousand years! Fortunately, nature provided us with ways in which this might actually be achieved.

Laser Sailing

Enter the laser, a word that originated as an acronym for light amplification by stimulated emission of radiation. It is a science-fiction—like invention of the 1950s that may provide an alternative to sunlight for providing the thrust a solar sail needs when it is far from the Sun. An ideal laser emits light at one wavelength, or color, and in a narrow beam. Unlike light emitted from the Sun or a light bulb, ideal laser light is highly directional and does not spread out in all directions—meaning that the inverse-square law does not apply. Such a laser can theoretically push our solar sail even when it is very far from the sun. Unfortunately, we cannot build an "ideal" laser, and even the best laser beam will spread out somewhat as it moves away from its source. This is due to a process called diffraction, which the interested technical reader can learn more about by referencing a good physics textbook. That said, a laser-driven sail is still an exciting possibility, as the diffraction-limit doesn't appreciably impact the performance of a sailcraft until distances much, much greater than those limiting solar sails are surpassed.

We discussed how sunlight could push a solar sail when the sailcraft is near the Sun and propel it outward into the solar system until it reaches approximately the orbit of Jupiter. We've also determined that we can use a laser to continue pushing the sail when sunlight is no longer available. So where do we put this laser, knowing that it will require a lot of energy to produce a beam powerful enough to cross the gulf of space and still provide the sailcraft with enough light for propulsion.

If the laser were built on Earth, the power problem would certainly be easily solved. Many industrial nations have a power infrastructure that could easily sustain the operation of the laser required for deep-space or interstellar flight. But Earth is not a good location, for many reasons. First, the laser would be travelling through our dense atmosphere, which would immediately produce not only significant degradation of the beam's intensity (lots of light would be lost during the beam's passage through the atmosphere), but it would also cause the beam to diverge, or spread out, much sooner than would otherwise be the case for a comparable beam generated in vacuum. Second, Earth rotates on its axis once every 24 hours. That means it would be impossible to point a laser toward a specific point in space for more than a few hours at a time and even this would require a complex pointing system as it would have to moving constantly to maintain its aim point as the Earth rotates. And don't forget that Earth is in orbit around the Sun, adding additional motion for which a pointing system must compensate. Lastly are the politics of basing the laser on Earth. If a country builds a laser powerful enough to propel a spacecraft through deep space, then it would have a laser powerful enough to knock down another country's aircraft, missiles, and even their orbiting satellites. Such a laser could be used as a weapon.

What about putting the laser in Earth orbit? The political problem would still remain. A large, powerful, space-based laser could threaten not only aircraft, missiles, and spacecraft, but anything on the ground (provided that Earth's atmosphere were transparent to the laser's wavelength). Pointing would also still be an issue. Recall that a spacecraft in orbit is not stationary—it is moving at very high speeds so that it can remain in orbit and not fall back to Earth. A craft in low Earth orbit (up to about 1000 kilometers), circles the globe approximately once every 90 minutes. Instead of sweeping across the sky once every 24 hours, the laser is now forced to do so every 1.5 hours! And the motion of Earth around the Sun is still a factor to be considered. Tracking and pointing may be much more difficult for an orbiting space-based laser than one located on the ground. What about power? Without a power grid to tap into, an Earth-orbiting laser would be required to generate its own power. Extremely large solar arrays or onboard nuclear reactors would be required to produce the energy needed to drive the laser. Though the atmosphere is no longer a problem, moving the laser from the surface of Earth to Earth orbit does not appear viable.

Where, then, can we find abundant power, no atmosphere to attenuate the beam, relatively stable pointing (so the laser can push on the sail for a long period of time with minimal active pointing required), and no fear of the laser being considered a military threat?

One option would be to place the laser in orbit around the Sun, as shown in Figure 10.1. If the laser station is relatively close to the Sun, then the inverse-square law works in our favor by making solar array panels capable of producing much more power. If we locate at one-half the Earth-to-Sun distance, the arrays will theoretically generate 4 × more power from the

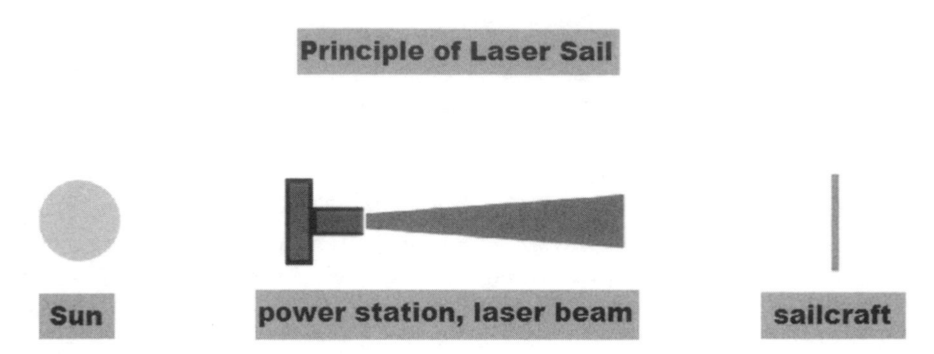

Figure 10.1. A laser-driven sailcraft could be accelerated significantly.

greater intensity sunlight falling upon them. There is no atmosphere, so there will be no immediate laser beam loss or added divergence. Pointing is still an issue, but it should be easier to steer and point the beam at our sailcraft from a laser in solar orbit because there is no longer planetary or planet-centric orbital motion that must be considered. Only the motion of the laser around the Sun must be accounted for. To compensate for the times when the laser is on the opposite side of the Sun from our sailcraft, two or three laser stations could be placed in solar orbit, with at least one of them always being in line-of-sight with the sailcraft, thereby providing propulsion.

A laser station closely orbiting the Sun is potentially not the best solution to the problem, however. Recall that we are concerned about not having enough light falling on the sail to allow it to continue thrusting once it passes Jupiter. If lasers didn't suffer from diffraction causing divergence, our problem would be solved. We could place our laser virtually anywhere and point it where we want it, without regard to distance. But lasers are diffraction-limited and they do diverge. Placing a laser close to the Sun only serves to reduce the maximum distance from Earth at which the laser light is still sufficiently intense to produce thrust on the sail. Ideally, we would place our laser at or near Jupiter so it can begin pushing the sail when the Sun completes its part of the job. Fortunately, Jupiter might be a great place for our laser.

At first glance, Jovian orbit seems to have nearly all the benefits of a solar-orbiting locale for basing the laser—except for power. The laser would not be located on a planetary surface, so there is not an atmosphere to contend with, nor is anyone nearby who might construe the laser to be a military threat. The motion of Jupiter around the Sun, and the commensurate viewing and pointing considerations, can be compensated for, as Jupiter orbits the Sun only once in 12 years. If the laser station is in a polar orbit around Jupiter, it could have a clear line-of-sight to our sailcraft for a decade at a time—taking into account only the orbit of the planet around the Sun. But what about power? Jupiter is far from the Sun, so solar power is not a good candidate. As discussed earlier, the laser station might be nuclear powered. Alternatively, the energy contained in the Jovian magnetic field might be harnessed with a long, conducting wire, or tether, deployed from the laser platform deep into the Jovian magnetosphere. The tether, due to its motion through the planet's magnetic field, would generate a potential difference across its length. This potential difference, or voltage drop, would result in the collection of electrons from the Jovian magnetosphere, thus producing a flow of electricity through the wire. The principle is the same as that which is seen when an electric generator produces electricity in a terrestrial power plant. On Earth, we produce electricity by moving wires through intense magnetic fields. Jupiter has the second most power magnetic field in the solar system, second only to the Sun. Our tether, moving through this field, can produce megawatts of power to drive the laser.

This is by no means the only scenario in which lasers might be used to push our sails. But it is certainly a likely one. A mission might proceed something like this: A sailcraft departs from Earth on a sunward bound trajectory. The craft falls toward the Sun and orients its sail to maximize solar thrust at perihelion, giving it an incredible boost toward the outer solar system. Sunlight continues to push on the sail until it reaches the orbit of Jupiter, at which point our tether-driven laser sends a beam of light to reflect from the sail, picking up from where the now-feeble sunlight leaves off. The laser maintains its aim point on the sail, providing continuous additional thrust, until the diffraction limit of the laser results in no net thrust being applied to the sail—somewhere in deep space. In this way, we can effectively extend the useful range of solar sails two- to fivefold.

Microwave Sailing

The laser is a powerful technology and it certainly represents one option to increase the effective range of a solar sail. But it is not ideal.

One problem of the laser is cost. Low-power lasers are fairly economical. One has to look no further than the ubiquitous compact disc or DVD player to realize that mass-produced lasers can be manufactured cheaply. Unfortunately, high-power lasers, with far-fewer commercial applications, are much more expensive to produce.

Happily, there is a far less expensive beamed-power alternative to the laser, although it too has its disadvantages. That alternative is the so-called "maser," or microwave laser. And high-power masers may be much less expensive to produce than their laser cousins.

There is no intrinsic reason why microwave-energy generators should be less expensive than lasers. The reasons for this cost disparity are tied in with military history and, as you might have guessed, mass production.

Large-scale generation of microwave power was pioneered during World War II by many of the belligerent powers. Radar, which uses microwaves, was developed in that era both to detect enemy aircraft at great distances by radar-beam reflection and to serve as a navigational aid. The enormous cost of developing high-power microwave generators was therefore born by the military establishments.

Besides cost, as with many technologies that appear magical at first glance, there is a catch. The wavelength of a microwave is generally in the millimeter-to-centimeter range. The wavelength of a near-infrared laser is about one ten-thousandth of a centimeter. By a mathematical principle called Rayleigh's criterion, the beam-spread or divergence of even a perfect laser depends upon the laser's wavelength.

You can get some idea of what this wavelength-dependent beam divergence means in practice by considering the following example. Let's say that you design an interstellar expedition to be accelerated by a near-infrared laser. To intercept all beamed energy at the extreme range of the laser, you estimate that the sail diameter is a large, but at least imaginable, 500 kilometers. But if you desire to save money on the propulsion mechanism and replace your laser with a 1-centimeter wavelength microwave transmitter of equal power and still have your sail intercept all transmitted radiation, your sail size must increase to a gargantuan five million kilometers, about three times the diameter of the Sun!

Clearly, something must be done or microwave-beamed propulsion becomes absurd. One possibility, as presented by Robert Forward, is to insert a thin-film focusing lens into the microwave beam between the transmitter and the sailcraft. Although such an approach, in principle, can deliver a lot more beamed energy to the sail, you must now contend with the problem of another large optical component that must be very accurately positioned in the depths of interstellar space.

During the early years of the 21st century, a NASA-funded team led by physicist Jim Benford and his author brother Greg Benford further investigated the problems and possibilities posed by microwave sailing. They concluded that microwave sailing might be best employed over short distances—such as accelerating a sailcraft from low Earth orbit to Earth-escape velocity using ground-transmitted microwave beams (a real possibility since the atmosphere is transparent to most microwaves, and existing radio telescopes can be used as transmitters).

The Benfords also employed a phenomenon called desorption that increases the efficiency of a microwave sailcraft. As well as pushing the sail by radiation pressure, microwave heating can evaporate gas molecules trapped in the sail during its manufacture, which can increase sail velocity. A small boost perhaps, but a boost nonetheless!

Particle-Beam Sail Propulsion

One disadvantage of radiation-pressure propulsion—of the solar, laser, or microwave variety—is the very small momentum of a photon. But what if we could construct a huge version of a nuclear accelerator to accelerate particles of matter to high velocity and impinge them against some form of sail? Just as in a solar sail, the reflected particles would impart some of their momentum and energy to the sail providing thrust.

Ground-based particle accelerators have been in use for decades in physics research. At Chicago's Fermilab, similarly to the other big particle accelerators in Europe and Japan, protons are accelerated in a many-mile-long accelerator and slammed into targets, or other accelerated beams of charged particles, to study the fundamental physics of matter. Charged particles like protons are used because we know how to make them move (accelerate). A charged particle in an electric or magnetic field will experience a force due to that field, making it move. By properly aligning the fields, these charged particles can be accelerated to very high speeds—close to the speed of light. If such a particle beam were to strike a sail in the vacuum of space, the sail would move and continue to gain speed as long as the beam impacts it.

Charged particle beams have one very serious flaw in their potential application to space travel—divergence. Unlike divergence in laser or maser sails, the divergence of a particle beam is caused by the accelerated particles themselves. The simple axiom, "like charges repel; opposite charges attract," dooms a charged particle beam sail from being useful at any significant distance from its source. As the beam of charged particles emerges from whatever accelerator created it, the very atoms within the beam, typically protons (with a positive charge), begin to push away from each other, until the beam spreads and becomes too diffuse to be useful.

In the heyday of the U.S. Strategic Defense Initiative (SDI), space-based particle beam weapons were being seriously considered as a method for shooting down or disabling missiles. To circumvent the beam-divergence problem inherent with their operation, engineers and scientists began developing neutral particle beam systems—neutral particles don't repel one another, thereby reducing or eliminating the problem of beam divergence.

The first step in producing a neutral particle beam is making a charged particle beam. Neutral atoms cannot be accelerated in an electric or magnetic field because they carry no net charge. Therefore, a beam of charged particles, typically protons, is first produced and accelerated to high velocities. Passing it through a very thin film or plasma cloud then neutralizes the beam. (To "neutralize" a proton means to simply provide it with an electron so that it becomes charge neutral, therefore not susceptible to charged-particle self-repulsion.) The charge-neutral beam can then propagate through space unimpeded to the target, or in our case, to the sail. As with most engineering solutions, the charge neutralization process is not without problems. It, too, induces beam divergence that causes the beam to spread out over long distances. This divergence is caused by the atoms of the beam colliding with atoms in the film or plasma cloud and reflecting from them as they "pick up" an electron.

Putting large, high-power neutral particle beam accelerators in space to propel starships may indeed be possible. We don't yet know how to engineer a system large enough, powerful enough, or with sufficiently low divergence, but there appears to be no physical reason we cannot. As with high-power lasers, the politics may prevent us from developing them: a high-power neutral particle beam system in Earth orbit could easily be used as a weapon.

Further Reading

Principles of beamed propulsion are reviewed in E. Mallove and G. Matloff, *The Starflight Handbook*, Wiley, NY, 1989. For a more up-to-date technical treatment and review, see G. L. Matloff, *Deep-Space Probes*, 2nd ed., Springer-Praxis, Chichester, UK, 2005.

AIP Conference Proceedings 664: First International symposium on Beamed-energy Propulsion, May 2003, American Institute of Physics.

Construction of Sailcraft

III

Designing a Solar Sail

For solar sails, as with most engineering challenges, there is no single, "best" design solution that will meet all potential needs and mission scenarios. This chapter discusses the most viable solar-sail design options and the pros and cons of each, and the problem of controlling the orientation of a sail in space.

Types of Solar Sails

Sail Physics Requires Some Design Commonality

Before we discuss the myriad options available to solar-sail designers, it might be useful to review some basics. First of all, a solar sail must contain a lightweight surface that efficiently reflects light. (At least until we discover a way to "virtually" reflect light, which is currently beyond the realm of realistic engineering possibility.) There is usually some sort of material under the reflector to provide structural strength and stability as well as to help balance any thermal issues. Current technology requires these lightweight materials to be deployed or suspended from some sort of boom, similar to the mast of a 17th century sailing ship or to spin, and have the deployment and deployed configuration maintained by the resultant centripetal acceleration. Building on these basic requirements, creative engineers and scientists have developed several options to consider as we begin solar sailing.

Three-Axis Stabilized Solar Sails

A three-axis stabilized solar sail most resembles a kite. Like a kite, booms support the solar sail material in three dimensions—the two dimensions that form the plane of the sail (left/right and top/bottom) as well as the dimension perpendicular to the plane of the sail (up/down). Like an airplane or a rocket, the sail must be stable in all three dimensions to allow the precise pointing required to control the Sun-provided thrust (pitch, roll, and yaw), thus allowing the sail to carry a payload where we want it to go. The sail must

G. Vulpetti et al., *Solar Sails*, DOI: 10.1007/978-0-387-68500-7_11,

Centripetal acceleration is the acceleration that causes any rectilinear path to become curved. It is a pure kinematical concept, which is not limited to circular motion. For instance, an object at the end of a rope, rotating about a vertical axis, undergoes a centripetal acceleration *caused* by the cord's tension acting toward the rotation axis. When one writes Mass $\times$ Centripetal Acceleration $=$ Tension, this means that the active force (or the motion cause) is the cord's tension, whereas the centripetal acceleration is the kinematical manifestation of this force. This is only a particular case of the general equation Force $=$ Mass $\times$ Acceleration.

Do not confuse *centripetal* acceleration with *centrifugal* acceleration, even though they have the same magnitude. The latter is sensed by a body in a rotating frame; for example, think of what you sense when you are steering your car along a highway curve. The centrifugal acceleration is directed outward with respect to the curve, as its name indicates. An observer on the highway (or on the other side of the police television circuit) has a different view by watching you and your car curving because of centripetal acceleration. In this case, such acceleration is the consequence of the car engine, wheels, and road–wheel friction.

In contrast, in a general rotating frame, any body undergoes three different accelerations (all independent of its mass) for the mere reason that it is rotating, namely, they are caused by no force. (Normally, the true explanation of that can be provided in a postgraduate course for physicists.) Here, with regard to solar sails, it suffices to mention that a body in a rotating structure senses (besides the centrifugal acceleration, which is proportional to the distance from the rotation axis), a second acceleration that depends on the body's relative speed (the Coriolis acceleration, very important also in air and ocean circulations). The third acceleration occurs when the frame rotates at nonconstant angular speed. (This last term may be included in a more general definition of centrifugal acceleration). The directions and the magnitudes of such accelerations differ from each other, in general.

also be supported in these dimensions to prevent it from going slack or collapsing on itself in any direction. Just imagine trying to fly a kite that has no supporting structure, and you will understand why a solar sail requires booms.

What characteristics must these booms have? First of all, given the overall size of a solar sail, typically greater than 20 to 40 meters on a side, and the relatively small size of today's rocket fairings (typically less than 5 meters in

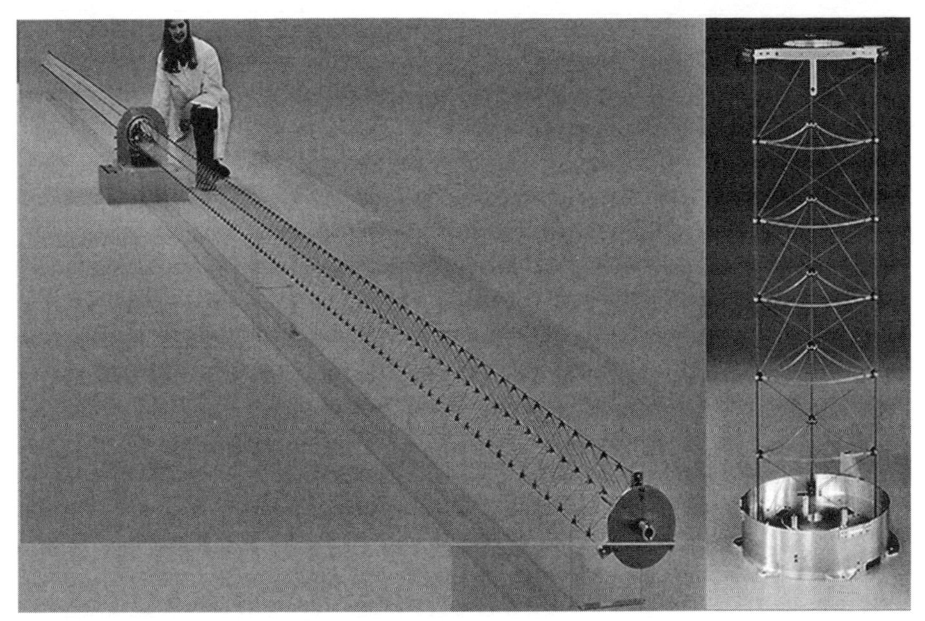

Figure 11.1. Capable of supporting a solar sail in space, this boom, developed by ATK Space Systems, was tested by NASA both in air and in space-like vacuum conditions. (Courtesy of ATK Space Systems)

diameter), the booms must be deployable from some sort of spacecraft. There is no rocket known that can loft a pre-deployed, 20-meter-diameter solar sail. These deployable booms must also be very lightweight. Recall that a key technology driver for a solar-sail propulsion system is (low) mass. The push from sunlight is slight, and if the sail or its support structures are heavy, the sail will not perform well.

NASA tested two design options for a long, lightweight deployable solar-sail boom in 2004 and 2005. The first option most resembled a sail ship's mast of days gone by in that it was a solid, mechanical boom. Made from state-of-the-art composite materials, a rigid mechanical boom (developed by ATK Space Systems of Goleta, California) was used to deploy and test a 20-meter solar sail developed for NASA. The boom and sail worked well in both ambient testing (room temperature and in air) as well as in thermal vacuum testing at NASA's Glenn Research Center Plum Brook Station (Sandusky, Ohio). The ATK booms, when stowed, resemble a spring under tension. They collapse to a mere 1 percent of their fully deployed length and, when deployed, are capable of suspending a large sail even under the effects of Earth's gravity, which they will not have to sustain during operation in space. Figure 11.1 is a picture of the mast during development testing by NASA and ATK.

Figure 11.2. This rigid boom was used by DLR to deploy its solar sail during ground testing in 1999. (Courtesy of DLR)

NASA's efforts in this area were preceded by Germany's Deutschen Zentrum fur Luft-und Raumfahrt (DLR), which used booms made from carbon fiber reinforced plastic to deploy a 20-meter 3-axis stabilized solar sail in 1999 (Fig. 11.2).

NASA also worked with L'Garde, Inc. (Tustin, California) to develop lightweight inflatable boom technology (Fig. 11.3). As the term implies, an inflatable boom is stowed onboard the central spacecraft structure until its deployment is initiated by blowing it up like a balloon. Nitrogen gas is expelled into the balloon-like boom until it is fully deployed. The boom is made from a material that quickly becomes rigid after exposure to the cold temperatures of deep space, thus obviating the need for the gas to remain within it. The benefits of this approach are twofold. First, the inflated boom is mechanically simple with few moving parts. Second, it is very lightweight and can be scaled to larger sizes without a significant increase in overall mass density. Since having low mass is critical for a solar-sail propulsion system, this approach holds much promise.

Spin-Stabilized "Solid" Solar Sails

An obvious question to ask when designing a lightweight solar sail is how do we reduce the amount of mass required. One answer is to eliminate the mass of the booms (described previously) used to deploy and stabilize sail. Fortunately, nature provides us with a proven and easily implemented

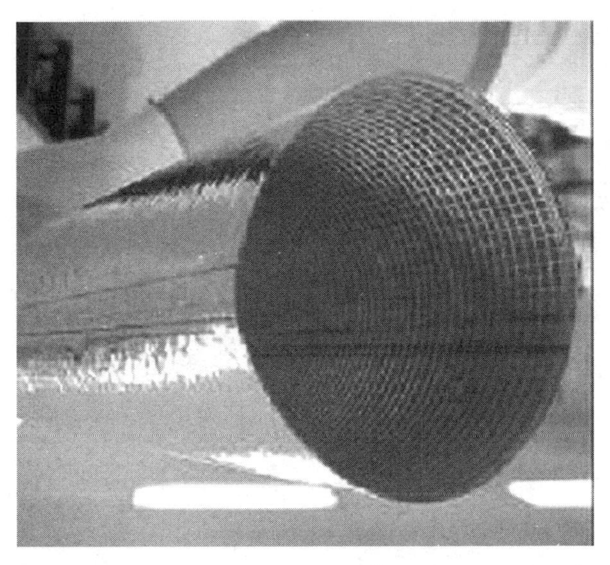

Figure 11.3. Shown here is deployment testing of L'Garde's inflatable boom for a solar-sail propulsion system. (Courtesy of NASA)

solution—we can spin the sail to get rid of the booms. The centrifugal acceleration experienced by the sail due to its rotation (as mentioned in the box above, the system of the sail's molecules, as with any rotating object, senses its own rotation point by point) puts the sail material under tension, keeping it flat as sunlight reflects from it, thus eliminating the need for any booms. This may require that the sail be strengthened with tension-bearing lines, but the mass required for these lines is much less than that of a boom system. Since the sail system is spinning, it behaves like a large gyroscope, providing stability in pointing that would otherwise have to be achieved in some other way.

In addition to providing pointing stability, keeping the sail flat, and under tension, a spinning sail can be easily deployed. The rotation acceleration that keeps the sail taut can be used to gently pull the sail outward from the spacecraft during deployment. One should note that, during the deployment process, the sail moves (slowly) with respect to the rotating structure. Such a deployment would work like this:

1. The sail is stowed aboard a small spacecraft and is launched into space.
2. The spacecraft begins to spin.
3. The folded or packaged sail is released from the spacecraft, slowly unfurling due to the centripetal acceleration produced by the spinning spacecraft.

Figure 11.4. The Russians were the first to deploy a spinning, solar sail–like structure in space. This is an artist's concept of Znamya-2. (Courtesy of Russian Space Agency) (See also color insert.)

4. The fully deployed sail is kept taut by maintaining a slow rotation about an axis perpendicular to the plane of the sail.

The Russians successfully demonstrated this technique in space with their Znamya mirror experiment flown in 1993. Deployed from an unmanned Progress spacecraft following its resupply of the Mir Space Station, a 20-meter circular sail-like reflector was unfurled. Its stated purpose was to demonstrate the technologies required to use large mirrors to illuminate cities at night, though most of the technologies on Znamya were directly applicable to solar sailing. A follow-up experiment in 1999 was to have deployed a 25-meter diameter sail from another Progress vehicle during space operations. Unfortunately, this test failed due to the accidental extension of an antenna into the area occupied by the unfurling sail—the antenna caused the sail to crumple, ending the experiment.

Spin-Stabilized "Heliogyro" Solar Sails

The heliogyro is another class of spin-stabilized solar sail. Heliogyro sails are also stabilized by centripetal acceleration, but they take on a totally different character in that they are composed of several separate vanes that deploy

Figure 11.5. Shown here is an artist's rendering of a heliogyro solar sail composed of multiple vanes deployed and stabilized by the spinning motion of the central spacecraft. (Courtesy of B. Diedrich)

because of the spinning motion of the centrally located spacecraft. Instead of appearing as a solid, or near-solid, circular reflector, they look more like a windmill. An artist's rendering of a heliogyro solar sail is shown in Figure 11.5.

This list is by no means exhaustive. In addition to the varieties of sails mentioned in this chapter, there have been various studies and technology efforts by the world's space agencies, universities, and private organizations that result in a myriad of design options. Some show the benefits of triangular 3-axis stabilized sails versus square ones. Others show the superiority of inflated booms over rigid mechanical booms, and vice versa. One thing is certain when comparing the various sail design options, no one option is superior for all mission applications or time frames. A near-term mission to study the Sun in the inner solar system will likely utilize a very different sail technology than that which will be used for our first missions into interstellar space. Engineers, keep innovating!

How To Maneuver A Sailcraft

What Is Spacecraft Attitude?

Let us begin by explaining spacecraft attitude. This concept is not limited to space vehicles or other bodies outside our planet. The classical astronomical observations of the celestial bodies have been done from ground by (automatically) projecting them onto a sphere of very large, but indeterminate, radius. Such a sphere is called the *celestial sphere*; it is a mental construct, but quite useful. This concept does not depend on the specific planet or other body one is considering. Thus, the direction of a star is simply the observer-to-star line. The intersection of such a line with the celestial sphere is a point that is completely determined by two angles, like the longitude and the latitude pair, at a given instant. In different words, a point on the celestial sphere represents a direction.

Now, let us suppose that you want to tell somebody how, for instance, a large hardcover book is oriented in your study room. The first thing to do is to define some frame of reference in your room walls. Recalling analytic geometry and the three Cartesian axes: x, y, and z, the frame of reference can be X-Y-Z along three edges convergent in one of the vertices (you may call the *origin* [O]) of the room. Because your book can take a lot of (infinite, in principle) orientations with respect to the frame O-XYZ, you can repeat the same logical process for the book. Thus, you have constructed another frame, say, o-xyz, *attached* to three edges emanating from a vertex of your book. (It is not mandatory that the two frames are Cartesian, in reality, but it's very useful.) At this point, the orientation of your book is quite determined once you decide the directions of the axes x-y-z with respect to the room frame XYZ. (Note that, as far as orientation is concerned, you don't need to know the position of the o-point with respect to the O-point). In practice, you have to know the angles that x-y-z form with X-Y-Z. Six of the nine possible angles are sufficient. Thus, you have determined the *attitude* of your book with respect to your room. If you rotate the book in some way, you can repeat the same steps for determining the new attitude.

In space, we don't have a pretty room for orienting a spacecraft. The role of your room can be replaced by the celestial sphere *centered* on the spacecraft. However, we need again three axes xyz bonded to the spacecraft's main structure; the origin of the axes may be coincident with the spacecraft's center of mass or some other suitable point. Quite similarly to the book example, the three directions of xyz represent the orientation or the attitude of the space vehicle.

Of course, the attitude of a spacecraft may change with time. The spacecraft can rotate about some axis of symmetry; thus, at a given time, one must also measure the rotation angle to get the complete attitude. The examples are manifold because any spacecraft may have rotating parts, flexible appendages, long booms, independent steerable pieces, damping internal systems, and so forth.

To specify angles on the celestial sphere, we need to define a *great* circle acting as a reference and a special point (E) on it. (A great circle on the sphere is a circle with its center coincident with the sphere center.) In turn, this reference plane defines its own *north and south poles*, namely, the intersections between the sphere and the orthogonal-to-circle line passing through the sphere center (C). Usually, the north pole (N) is adopted as the second reference point. Thus, the CE line is taken as the x-axis, whereas the line CN is taken as the z-axis. Hence, the y-axis is automatically fixed. The two special points, E and N, are utilized to measure the angles defining a direction. Historically, the great circle of reference was Earth's equator at some date and the E-point was the east intersection of the ecliptic with the equator (the March equinox). Nowadays, the equatorial system of coordinates has been replaced by the highly-accurate frame known as the International Celestial Reference Frame (ICRF, or its idealization ICRS), which is strictly inertial. The ICRS orientation, though, is close to that of the old system taken at J2000 (this abbreviation stands for the date 2000-01-01, 12:00:00, terrestrial time). The interested reader can be introduced to or find technical readings on such basic topics at http://www.iers.org/.

Classifying Attitude Analysis Items

The general spacecraft attitude analysis may be categorized mainly as attitude *determination*, attitude *prediction* and attitude *control*.

Attitude *determination* is the process of computing the spacecraft orientation, with respect to an inertial frame of reference of Earth, the Sun, or another celestial body, starting from the measurements of sensors onboard the spacecraft.

Attitude *prediction* consists of forecasting the future evolution of spacecraft orientation via algorithms, where both the spacecraft and the environment are modeled.

Attitude *control* is the process that enables us to get the desired attitude in a certain period of time for different purposes (e.g., thrust activation, spacecraft safety, scientific payload requirements, perturbation compensation, and so forth). There are two main areas: attitude *stabilization* and attitude *maneuver*. The former concerns a process aiming at keeping the spacecraft attitude for a certain time interval. The latter concerns the problem of changing the spacecraft attitude, especially for allowing the spacecraft to follow the right trajectory to the mission target.

Finally, any spacecraft may be categorized in two large classes with respect to the attitude stabilization: (1) spin-stabilized spacecraft, and (2) three-axis stabilized spacecraft. The second class requires more complex active control of the vehicle attitude, which otherwise would drift uncontrolled under the action of external torques that may continuously perturb the spacecraft.

Classically, celestial mechanics is the area of dynamics and astronomy that addresses the motion of celestial bodies under their reciprocal gravitational influence. Astrodynamics is the study of the motion of artificial objects in space. The big difference between astrodynamics and celestial mechanics consists of propulsion and its control. In turn, astrodynamics has two major partitions: trajectory (or orbit) dynamics, and attitude dynamics. The former addresses the motion *of* the center of mass, or the barycenter, of spacecraft (i.e., the translational motion), whereas the latter is concerned with the motion of the spacecraft *about* its barycenter (i.e., the rotational motion).

When a force (either internal or external) applies along a direction that does not pass through the barycenter, the so-called moment of the force or the *torque* (about the barycenter) is generated. Internal and external torques can affect the rotational motion of parts of spacecraft with respect to others. However, only the external torques act upon the *overall* rotational motion (e.g., with respect to an inertial frame) of the spacecraft.

A fundamental property of spacecraft motion is that its trajectory and attitude histories are strongly connected. Conventionally, propulsive devices for trajectory control are called the main engines or thrusters, whereas the devices providing spacecraft with the control torques for orientation maneuvers are often referred as the control hardware or the *actuators* (which therefore represent a component of the whole attitude control system).

Sail Attitude Control Methods

In general, there are two major external torques on any spacecraft: (1) the *disturbance torques*, caused by the space environment the spacecraft interacts with; and (2) the *control torques*, induced intentionally by means of attitude actuators. The latter are of utmost importance because it is through attitude evolution that the main propulsion system, whatever it may be, forces the spacecraft to follow the planned trajectory to the final target.

A general rigid body rotating freely (no torque) in space has a rather complicated motion, with angular velocity constant in magnitude, but variable in direction (i.e., something roughly like the uniform circular motion). If we want to change both the magnitude and the direction of the angular velocity, and then to affect the attitude angles, we have to apply torques.

In Chapters 5 and 7 we stated that two points, normally inside the volume occupied by the whole sailcraft, are given special importance in sailcraft dynamics: the center of mass of the spacecraft and the sail system, and the center of pressure of the sail system. Since the sail system is much larger than the spacecraft, one can define the vector position of the spacecraft (as a point-like system) with respect to the sail. In addition, the solar-pressure thrust vector has a major component along the sail axis and a (nonnegligible) component along the mean plane of the sail (see Chapter 16 for more precise explanations). Normally, the sail axis is not aligned with the (local) Sun–sail line and there is the need to change the attitude sail systematically for controlling the sailcraft trajectory. Let us describe some methods envisaged for attitude control.

Method 1: Relative Displacement Between Barycenter and Center of Pressure

One can think of shifting the sail laterally by acting on the sail structure directly. This sail shifting should be easier to implement if the sail were like a one-block structure. If the full sail is sectioned into subsails or panels, then one or two symmetrical sections may be translated with respect to the others. In any case, a torque arises with respect to the barycenter. Such torque can affect only two of the three sail directions; it is not possible to control the motion of the sail about its orthogonal axis. That may cause problems to the attitude control of some scientific instruments of the spacecraft payload, if a three-axis control is required by the mission objectives.

Since it is the relative displacement that matters, one could shift a ballast mass (in the spacecraft) by some device consuming electric energy. The physics of control does not change, of course: the induced torque allows two-axis control, as above. However, the towing device may be much simpler and lighter, especially when the sail is very large.

The so-called control authority is strongly related to the sail attitude itself; in other words, if the sunlight impinges on the sail with a large incidence angle, not only does the thrust acceleration decrease, but also the torque that one wants to use for controlling the sail lessens. Furthermore, the spacecraft has to be located between the Sun and the sail, a constraint that would cause many problems in missions for which the sailcraft trajectory is close to the Sun. There is another general risk. A number of non-ideal effects may induce the barycenter and the center of pressure to be offset by some unwanted (and unmeasured) position vector: hence, emerges *bias or unbalanced force moments*, which act as disturbance torques. As they may be comparable to the attitude-maneuver-required torques, they need to be trimmed down to zero (nominally); this can be done by an active attitude stabilization device, namely, by increasing the mass and complexity of the sailcraft.

Method 2: Using Pairs of a Segmented Sail

In contrast to the above-mentioned subsail pair (which was translated), this technique would use the rotation of two opposite panels. To do so, the sail has to have each attitude panel supported by two articulated booms, which gimbal at the sail mast structure. In addition, at the boom tips, the panels may be attached to small movable spars; thus, panel edges can be independently raised or lowered with respect to the boom's plane. By such a method, full sail controllability can be achieved. However, the hardware that enables panel movements should be rather massive. For redundancy, at least two panel pairs have to be equipped as described, thus increasing the sailcraft mass to sail-area ratio. As a result, the solar-pressure thrust decreases, and fast missions would not be allowed.

However, for other mission types, this control technique exhibits two additional advantages: (1) Attitude control is still possible when, in some mission phases, the sunlight is grazing the sail's mean plane, namely, when thrust is almost zero, (2) A priori, the spacecraft is not constrained to be put between the Sun and the sail, unless otherwise required.

Method 3: Utilizing Small Sails Located at the Boom Ends

In recent years, this method has been investigated considerably with regard to four-panel squared sails, like those experimented on the ground by NASA and ESA. Figure 11.6 visualizes the concept for a squared sail typically assumed for the first missions. One attaches small sails, or vanes, at the end of the booms of the mainsail frame. Each vane is a complex structure quite similar to the sail system. A vane may have either a triangular or rectangular shape. Every vane frame is gimbaled to the boom tips in such a way as to have one or two independent rotational movements. Thus, a full control authority of the mainsail could be achieved, even for real sails with construction asymmetries, beam bending, and billowing. Each pair of opposite vanes can be given a different task; for example, following Figure 11.6, the red pair of vanes (the *fore* and *aft* vanes) is more or less aligned with the sailcraft velocity. It can be used for getting and stabilizing a desired value of the angle between the sail's normal (the ideal sail axis) and the sunlight direction. The green pair of vanes (the *starboard* and *port* vanes) can be utilized for maneuvering and performing an active stabilization about the

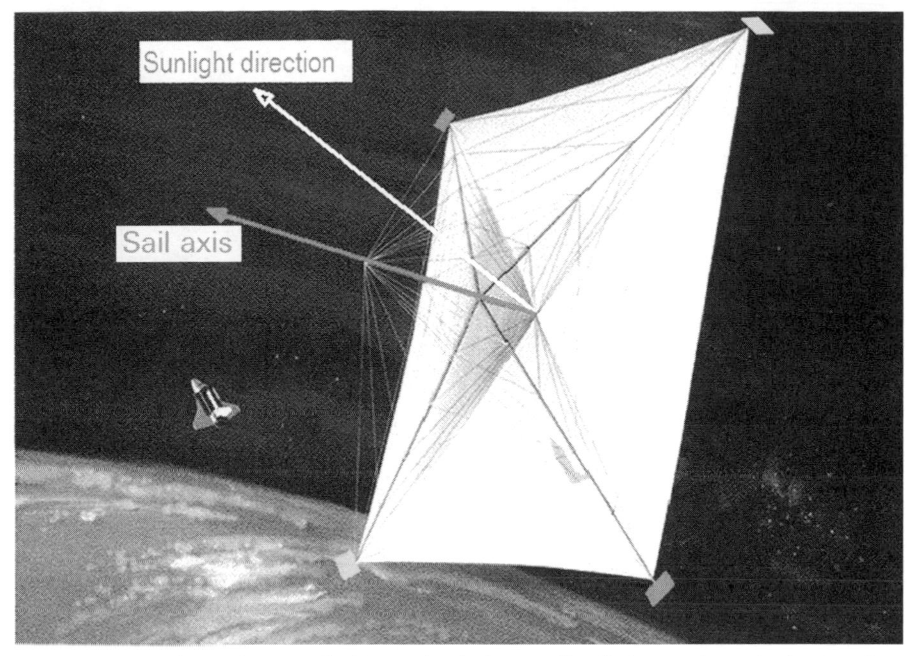

Figure 11.6. Solar sail controlled in attitude by small sails located on the boom tips. (Courtesy of NASA, adapted by G. Vulpetti) (See also color insert.)

current sunlight direction. By sequencing the two maneuvers in either order, one can get the desired attitude of the sail and stabilize it for a certain time, until a new attitude is required.

This technique of sail attitude control seems difficult to apply to circular sails. The circular "rigidized" beam that keeps the sail open (cf. Chapter 7) would be too small for a strong joining to complex structures like vanes.

Method 4: Very Small Rockets

This is an obvious and well-known technique. Depending on the mission duration and goals, one may employ microchemical engines or microelectric thrusters. On the balance scale pans, one has two main conflicting "weights": the consumption of propellant and the independence of the sailcraft distance from the Sun. As a point of fact, the previously mentioned methods utilize the solar pressure, which acts on the attitude control surfaces as well. However, in missions to distant planets, a spinning sail is not appropriate for (long) rendezvous maneuvers. Thus, full sail control would be necessary, but this gets complicated because of the weakness of the solar pressure with the increasing Sun–sail distance. On the other side, using micro-rockets as primary devices for attitude control for the whole mission may result in an unfavorable mass, especially for large-sail missions.

Conclusion

The first solar-sail missions in the near term, in particular the flights of technology demonstration, may use one of the techniques described above. Subsequently, as experience accumulates and the mission complexity increases in terms of goals, transfer trajectory, and operational orbit, a multiple attitude control system may turn out to be the most efficient choice. For instance, methods 1 and 3 would entail an excellent propellant-less control, while method 4 (e.g., via pulsed plasma-jet micro-rockets at the sail mast tips) guarantees a backup attitude subsystem independent of the solar pressure. In this case, all three methods would make up the full attitude control (and stabilization) system.

The above methods represent classical techniques, in a certain sense. In Chapter 12 on emerging technologies, we discuss some unconventional ways to control the attitude of a sailcraft.

Building a Sailcraft 12

Using Today's Technologies

Despite its decades old theoretical foundation and all the efforts of space researchers, there have been surprisingly few attempts to build and fly large solar sails in space. Germany's DLR took particular interest in solar sail technology in the 1990s and fabricated one of the first large-scale ground-based engineering model sails. The Russians demonstrated in space a spinning 20-meter mirror, called Znamya, from their Progress resupply vehicle after it completed its mission to the Mir space station in 1993. Though technically a mirror, the technologies used were essentially the same that would be required to build a solar sail. The Planetary Society, working with the Russians, developed a sail and would have demonstrated the technology in space had the rocket not failed during a launch attempt in 2005. The Japanese are also developing solar sails. In August 2004, an S-310 suborbital rocket launched from the Uchinoura Space Center in Kagoshima, Japan, deployed 2 types of solar sail materials to validate both the materials and their deployment in space. The Japanese sail experiment was a success, though it was not a demonstration of a free-flying solar sail that could be used for deep-space exploration. In 2005, NASA built and tested on the ground two 20-meter solar sails, each using very different technical approaches.

Russia's Space Mirrors

Though technically a demonstration of space-based mirror technologies capable of reflecting sunlight from space to illuminate cities after dark, the Russian Znamya experiments were actually the first space demonstrations of solar sails. In 1993, Russia launched a Progress vehicle to send supplies to their Mir space station. After completion of the mission, the Progress undocked from the Mir and took up position some distance away. Russian

G. Vulpetti et al., *Solar Sails*, DOI: 10.1007/978-0-387-68500-7_12,

cosmonauts aboard Mir then commanded the vehicle to begin spinning a canister containing the sail structure. Using the resulting centripetal acceleration, the sail/mirror deployed outward, forming a 20-meter-diameter thin-film reflector. The mirror/sail was monitored and then jettisoned to burn up in the atmosphere. An unsuccessful follow-up Znamya experiment was launched in 1999. This test failed when an antenna became entangled with the mirror/sail during its unfurlment.

Germany Advances Sail Technology in the 1990s

In December 1999, Deutschen Zentrum fur Luft-und Raumfahrt (DLR) and its partners, the European Space Agency (ESA) and INVENT GmbH, deployed a 20 meter by 20 meter square, 3-axis stabilized solar sail in a ground test facility (Fig. 12.1), thus proving their sail fabrication, storage and deployment capabilities. The effort was part of the DLR-led consortium's proposed ODISSEE (Orbital Demonstration of an Innovative, Solar Sail–driven Expandable structure Experiment) mission. Alas, the proposal effort was not successful and ODISSEE was not selected for flight.

The DLR sail booms were made of a carbon fiber reinforced plastic. Each boom consisted of two laminated sheets, which were formed into a tubular shape that could be readily flattened around the central, payload-carrying

Figure 12.1. A photograph of the DLR solar sail ground demonstrator under full deployment. (Courtesy of DLR) (See also color insert.)

portion of the spacecraft. During deployment, the booms uncoiled into their tubular shape and served as the mast and point of attachment for the sail material. The sail was made from aluminized mylar. The sail was pulled from the central structure in much the same way that sailing ships unfurl their sails—with a network of wires and ropes, many of which were driven by an electrical motor.

Cosmos 1: The Planetary Society Put Its Money on the Table

If all had gone as planned, the U.S.-based Planetary Society, working with Russia, would have been the first to fly a fully functional, though performance limited, solar sail in space. The project, called Cosmos 1, was financed with the private contributions of space enthusiasts from all over the world. Once in orbit, the sail was to have deployed using inflatable booms and a set of 8 triangular blades in the "heliogyro" configuration (Fig. 12.2). Had it been successful, it would have been the first spin-stabilized, free-flying solar sail to fly in space. Unfortunately, the Russian rocket, a converted submarine-launched ballistic missile (SLBM), malfunctioned and did not place the sail spacecraft into orbit.

Figure 12.2. Sketch of The Planetary Society's Cosmos 1. (Courtesy of The Planetary Society.) (See also color insert.)

Figure 12.3. The L'Garde solar sail before vacuum testing at NASA's Plum Brook facility. The scale of the device can be appreciated by examining the relative size of the people in front of it. (Courtesy of NASA) (See also color insert.)

NASA Gets Serious About Solar Sails

NASA worked in earnest to develop a credible solar sail technology between 2001 and 2005. During this time period, two different 3-axis stabilized, 20-meter solar sail systems were developed and successfully tested under thermal vacuum conditions. The two competing sails were designed and developed by ATK Space Systems and L'Garde, Inc., respectively. Both sails consisted of a central structure with four deployable booms that supported the sails. These sail designs are robust enough for deployment in a one-atmosphere, one-gravity environment, and are scalable to much larger solar sails, perhaps as much as 150 meters on each side.

L'Garde Inc., of Tustin, California, developed a solar sail system that employs inflatable booms that are flexible at ambient temperatures but "rigidize" at low temperatures. Their concept uses articulated vanes located at the corners of the square to control the solar sail attitude and thrust direction. L'Garde's technology uses the same sail material as the DLR sail—aluminized mylar (Fig. 12.3).

ATK Space Systems (formerly Able Engineering), of Goleta, California, developed a coilable longeron boom system that deploys in much the same way a screw is rotated to remove it from an object. Attitude control is achieved with sliding ballast masses to offset the center of mass from the center of pressure of the sail. Roll control is achieved using spreader bars at the tips of the mast, which causes the sail to behave much like a pinwheel. Instead of mylar, the Able Engineering team used CP-1, a proprietary material provided by SRS Technologies.

Both hardware vendors fabricated and tested 10-meter subscale solar sails in the spring of 2004. In 2005 they conducted 20-meter subscale solar sail deployments in vacuum at the NASA Glenn Research Center's Plum Brook Station in Sandusky, Ohio.

Japan Sails in Space

It is clear that the Japanese are taking solar sail technology seriously and are moving forward with missions to demonstrate its feasibility. Their first space test occurred in 2004 with the launch of a suborbital S-310 rocket that deployed candidate sail materials in a short-duration experiment that by all appearances was quite successful. Another sail test occurred in early 2006, (see Chapter 18 for additional information).

Current Solar-Sail Technology: Where We Stand Today

In the near term, it is evident that we will be limited in our selection of materials and structures for solar-sail missions. With the use of composite booms and mylar sails, the best possible sail areal density is certainly no less than approximately 10 grams per square meter (g/m^2). This will limit first-generation sails to be less than 150 to 200 meters in diameter and restrict the amount of payload that can be carried. Even with these restrictions, multiple science and exploration missions will be achievable with solar sails that are otherwise currently beyond our technological capabilities. Where do we go from here? Are there new materials that might get us down to an areal density of 1 g/m^2 or less? The answer is a cautious "maybe"!

First of all, based on the results of NASA's ground tests of both the ATK and L'Garde solar-sail designs, analysis indicates that both systems can achieve a loaded (loaded means that the spacecraft and scientific payload are included in the analysis) areal density of 10 g/m^2 or less. The L'Garde inflatable boom system appears to be scalable to lower areal densities, perhaps as low as 5 to 7 g/m^2. If so, then multiple inner solar system missions are now within our reach.

To fly more ambitious missions, such as the interstellar probe, will require solar sails with areal densities as low as 1 g/m^2. Clearly, this cannot be achieved with any of the materials or sail systems demonstrated so far, and a new approach must be developed. Fortunately, innovative people and companies around the world are already working on the problem and have some interesting and technically sound approaches to its solution.

A solar-sail breakthrough occurred in the early 1990s with the development of a carbon fiber sail substrate by Energy Science Laboratories, Inc. of San Diego, California. The substrate, woven into a mesh, consists of a network of carbon fibers cross-linked together. The fibers are very lightweight and ultra-strong, allowing a mesh (which is mostly empty space) to be constructed. The material is rigid, thus obviating the need for as much support structure as is required for current technology sails. It also retains its strength and other properties at high temperatures, a key requirement for solar-sail missions that are required to operate very near the Sun.

CU Aerospace is laying the groundwork for an alternative, ultra-lightweight solar-sail system called "UltraSail." As described in a paper presented at the July 2005 American Institute of Aeronautics and Astronautics Joint Propulsion Conference, the UltraSail system includes

> a central hub where the payload would reside. Attached to this hub would be several "blades" of solar sail film material that would unroll from a storage mandrel with the help of a tip microsatellite that is attached to the end of each blade. The baseline UltraSail design has four blades composed of a micro-thick reflection-coated polyimide film. During the deployment of the blades, the formation flying tip satellites spin up the entire system to create a spin-stabilized, controllable solar sail system with a large sail area.

The primary payoff of the UltraSail would be the elimination of the truss structures inherent in most of the sail systems proposed and tested to date. This would significantly reduce the overall areal density, allowing gossamer sails of 1 g/m^2 to be fielded. As humanity's space-faring technology and in-space infrastructure mature, we can expect many improvements in sail design and construction. Some of these improvements will be due to enhanced capabilities to deposit thin films in the high-vacuum, microgravity environment. Not to be ignored is the eventual possibility of constructing solar photon sails and associated equipment from materials found on the Moon and near asteroids, reducing the mass required to be launched from Earth. Finally, terrestrial materials technology will certainly improve, resulting in reduced sail areal density, stronger sails and cables, and high temperature-resistant sail materials.

One improvement might occur within a decade or so of the first operational solar photon-sail missions. First-generation sails are generally tri-layered. Facing the sun is a reflective layer, which is affixed to a plastic substrate. Among other things, the plastic substrate provides the structural strength required for the sail to survive the accelerations experienced during launch. On the anti-sunward side is an emissive layer that radiates the small fraction of incident sunlight that is absorbed by rather than reflected from the reflective layer. Some researchers have already conducted experiments with plastics that will evaporate when exposed to solar ultraviolet radiation. If this evaporation can be controlled in a large, thin-film structure, the areal density of Earth-launched sails will be greatly reduced.

Farther in the future, we may not only mine celestial bodies for solar-sail raw materials; we may also utilize the dynamic properties of these objects. Consider, for example, the possibility of unfurling a Sun-sensitive plastic film on the "night side" of an asteroid or comet. Then, using the process of vacuum-phase deposition, which allows the deposition atom by atom of nanometer-thin metallic films, a metallic reflective layer is built up on the plastic substrate. Finally, the completed structure is maneuvered into sunlight. The plastic layer evaporates, leaving only a hyperthin sail, which is now ready to roll. Furthermore, even the hyperthin metallic monolayer sail might be superseded. There are mass advantages to the perforated sail, in which perforations are smaller than a wavelength of light.

Nevertheless, the concepts behind the above potentialities belong, in a certain sense, to past physics and to the ways of using it. One could wonder, What about Tomorrow? Might there be any scientifically imaginable turning point that renders space sailing much more attractive than is now conceivable? In the previous chapters, we have dealt with a few pieces of such topics; in the next two sections, we discuss realistic scientific answers.

Using Emerging Technologies

Getting straight to the point, *emerging technology* means *nanotechnology*. What is nanotechnology? In most textbooks and introductory papers, you'll read that it is difficult to define nanotechnology. That's true. However, to explain things in a simple way and with our eyes toward the application of concern in this book, let us start with what is indeed at the base of nanotechnology, namely, *nanoscience*. Strictly speaking, *nano* means one billionth of something; here, the same word means one nanometer, or 1 nm, which is about ten times the size of a hydrogen atom. However, the scientific investigations use a scale roughly from 1 to 100 nm. In the upper part of this range, say, 60 to 100 nm, classical physics still holds for objects of such sizes, even though single components are driven by quantum physics. As an example, the new computer central processing units (CPUs) put on the market in 2006–2007 have been characterized by the 65-nm technology based on photolithography (lithography is a technique for printing onto a flat surface that was invented by Austrian actor Johann Alois Senefelder in 1798).

In the lower part of the above range, say, 1 to 30 nm, quantum physics dominates and new phenomena are expected. Perhaps more interesting is the zone of roughly 30 to 60 nm, where a sort of hybrid or transition between classical and quantum physics occurs. (Many current interpretations of quantum mechanics may be invalidated when the transition zone is understood.)

Nanoscience may be defined as the investigation of the fundamental principles of molecules and atomic/molecular structures with *at least one* dimension in the 1 to 100-nm range. In general, these structures are the *nanostructures*, which are complex systems in which some of the known physical laws might break down. They can be differentiated in terms of the number of their dimensions; for example, two-dimensional nanosurfaces can have thickness in the 1 to 100-nm range, whereas nanoparticles are three-dimensional with a radius of some tens of nanometers, at least. *Nanotechnology* is the application of these nanostructures into nano-scale devices, which would be designed for accomplishing *specific* tasks.

The attentive reader may be getting somewhat confused. In the 20th century, we saw incredible advances in atomic physics, nuclear and particle physics, microbiology, chemistry, science of materials, and so on, with plenty of applications, which have been pervading our everyday life. Some levels of nanotechnology are present today. So what is the difference between the current (micro and submicro) technologies and the many branches of the *future* nanotechnology? Well, embedded in the concept of nanotech there is a magic word: *control*. Controlling what? Nanotech aims at manipulating single atoms and molecules, for example, catching a certain molecule (by means of a tiny device), moving it, and releasing it at a certain place; also reaching certain molecules in a single larger structure and inducing the desired chemical reactions. In every human activity, examples of nanotech application are so many that they are limited only by our imagination!

When the number of atoms and molecules to be dealt with one at a time is very high, the time necessary to build the desired devices or to accomplish jobs can be so long as to be quite impractical. Then, programmable molecular machines, called *assemblers* by K.E. Drexler, shall be used. Assemblers are the nano-scale counterparts of the current industrial robots, and would be the high-end products of nanorobotics. (Nanorobots, or *nanobots*, are more general terms including 0.1 to 10-micron devices that can have nanoscale interactions with generic tiny objects. As of 2006, no nonbiological artificial nanobots have been realized.) One of the main features of an assembler is self-replication. The principle is as follows: if an assembler is able to accomplish any specific (nanoscale) task (even several times), then it should be possible to build an assembler capable of copying it, with exactly the same structure and the same program. There is a dense literature speculating on the pros and cons of the self-replication power.

The technical reader may remain perplexed about the general objectives of nanotechnology and could point out problems not with some practical feasibility like the industrial investments and costs, but with the conceptual fundamentals. The reader could argue that, since every structure is endowed with a temperature above absolute zero, atoms and molecules would not be controlled as we want because they undergo thermal noise (molecules move, vibrate, and rotate). Now it is possible to prove mathematically that the squared error in positioning atoms or molecules is proportional to their temperature (as expected), but inversely proportional to their *stiffness*. In general, the stiffness of an object is the ratio between a force and the deflection caused to it. Thus, because it is not always possible to decrease the temperature of the system hosting the molecules to be moved and positioned, either nanorobots or assemblers have to apply sufficiently intense forces to the target molecules to reduce the positional error to the desired values, even 1 nm or less. The technical reader may object about something deeper: What about quantum effects on the nanoscale processes we want to control? This time the answer is still more important and favorable! Equations governing the positional error are a bit more complicated than the previous case, but there is a key result: the more massive the molecule, the lower the positional uncertainty. For instance, molecules of normal sugar, a relatively simple type of carbohydrate, could be positioned with a precision of 0.01 nm!

All this is not surprising because, for hundreds of millions of years (at least), *biological* molecular machines exist and continue to work very well. This experimental evidence on a planet where life springs up is fundamental in concluding that artificially controlling atoms and molecules—even singly—is possible. In a certain sense, nanoscience and technology are favored with respect to other scientific areas for the mere reason that natural tiny machines can be mimicked efficiently. Of course, we have to learn many things and, above all, use them on behalf of the humankind, not against it.

Nanoscale objects have to be seen in a scientific sense, namely, detected, analyzed, and measured. There are many different methods and tools for experimental investigation, which received a big stimulus in recent decades. The main methods can be summarized as microscopy, spectrometry, and diffraction. In addition to electron microscopes, particularly efficacious are the atomic force microscope (AFM) and the scanning tunneling microscope (STM). Both instruments can produce high-resolution three-dimensional images of single atoms on a surface. Future versions of such microscopes could utilize nano-objects to increase their performance.

To describe the current areas of scientific and technological investigation, even very generally, is beyond the scope of this book. However, we will mention those that could have a strong impact on solar sailing. Such influence concerns the large sailcraft (obviously) and the so-called nanosailcraft (less obviously). Before inducing any misunderstanding, a nanosailcraft is not a spacecraft with a sail system 1 to 100 nm in diameter. The reference quantity is the sail area of 1 square kilometer (km^2). Why? A hypothetical (advanced) sailcraft with a sail loading of 1 g/m^2 and a sail of 1 km^2 may transport a (high indeed) scientific payload of 300 kg well beyond the solar system with a cruise speed very close to 0.001 c (or 63 AU/year)! We define the nanosailcraft as a sailcraft of total sail loading lower than 1 g/m^2 and with the sail area in the 0.001 to 0.01-m^2 (10–100 cm^2) range. Such a nanocraft may be one of several components of a sailcraft fleet (or a swarm), a concept we briefly mentioned in Chapter 7.

For the sail system, what one essentially expects from nanotechnology is essentially a quality leap from using special materials. One would need ultra-thin and high-strength (at the same time) sails, possibly mono-layered in order to avoid chemical/physical problems between different layers in the highly variable space environment. Sail and supporting structure materials should be of very low density and resistant to temperatures much higher than that (600 K) of the today-envisaged all-metal (Al-Cr) sail. Because we are not yet happy about all such properties, we also require that such particular materials undergo low degradation in their thermo-optical properties (caused by ultraviolet and extreme ultraviolet photons, and by the particles of the solar wind). As a result, an interplanetary sailcraft may last a long time in space: for instance, think of a large sailcraft working as a highly reliable shuttle between Earth and Mars for many years.

What about the capability of getting self-repairing sails? In addition, considering what we said about the sail attitude control methods in Chapter 11, one may wish to control a sail by acting directly on its optical properties! Science fiction, perhaps? No. One of the best potentials for all of that comes from carbon's third allotrope, discovered in 1985, named the *fullerene* or, better, the fullerenes, since they represent a set of carbon allotropes. Thus, besides diamond and graphite, carbon exhibits other chemical forms such as the fullerenes.

Figure 12.4 (left) shows the computer image of the basic molecule (buckyball) of fullerene, chemically indicated by C_{60}. It was discovered by

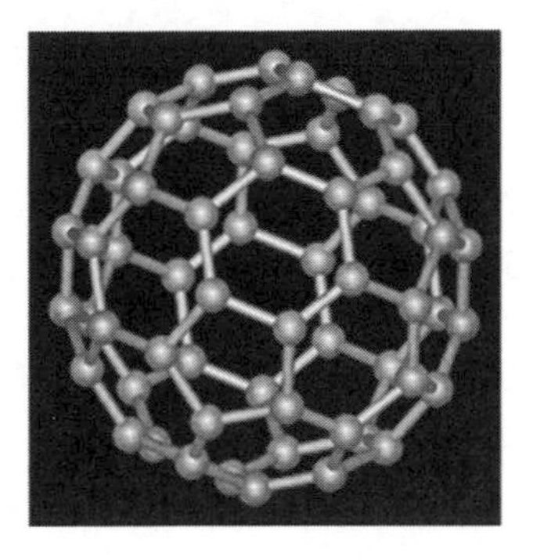

Figure 12.4. Molecules of fullerene.

R. Smalley, H. Kroto, R. Curl, and a number of graduate students at Rice University (Houston, Texas) in August 1985. C_{60} consists of 60 carbon atoms arranged like a soccer ball. Technically, the solid shape is named the truncated icosahedron. It has 32 faces: 20 are regular hexagons and 12 are regular pentagons. At each of the 60 vertices, there is a carbon atom. The diameter of such a quasi-sphere is of the order of few nanometers. Buckyball could be considered a quantum dot of zero dimensions, in practice. As said above, there are several fullerene types; for instance, the right of Figure 12.4 shows the structure of a 540-atom molecule. Although such molecules have striking properties, things do not end here. The top of Figure 12.5 shows a carbon layer 1 atom thick. These are the sheets that graphite is composed of; such a flat structure is called a *graphene*. There are several variants of graphene, but graphene is *not* an allotrope of carbon. Instead, graphene is unstable with respect to curvature. Thus, *conceptually speaking*, a graphene sheet could be rolled, one or more times, producing carbon *nanotubes* (shown at the bottom of Fig. 12.5). However, the actual methods of nanotube production are quite different from graphene rolling. Depending on their lattice structure, three main types of nanotubes arise: *armchair* (metallic nanotube), *zigzag*, and *chiral* (semiconducting nanotubes). Nanotubes may be arbitrarily long (though currently limited by laboratory and industrial devices), but their diameter can be slightly more than 1 nm. As Figure 12.5 shows, they are quite empty internally; as a result, one hypothetical nanotube from Earth to the Moon would weight about 0.3 milligrams! Why have nanotubes been increasing in importance since 1991, when NEC's

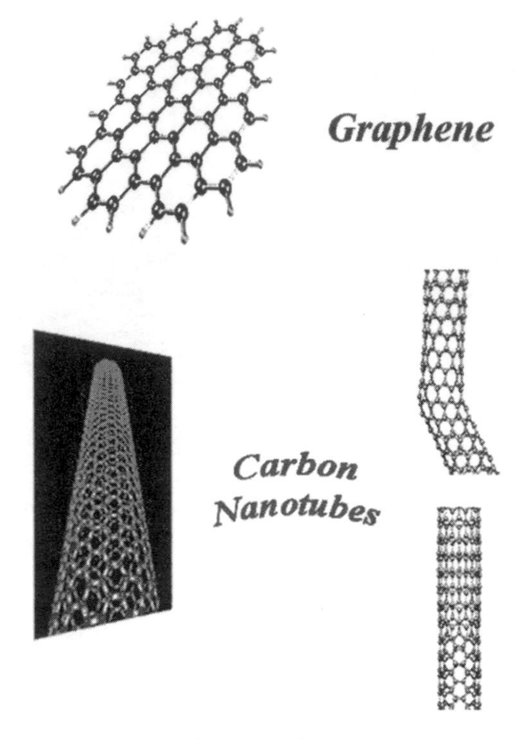

Figure 12.5. Carbon's flat sheet (graphene) and nanotubes.

researcher Sumio Iijima manipulated some of them by demonstrating their nature? (Previously, other researchers showed images of tiny tubes, but nothing more, in practice.) In addition to incredible values of the many classical mechanical (outperforming the best steel), electric, and thermal parameters, they exhibit quite unusual properties. For instance, electrons can travel along a carbon nanotube with energy-independent speed, namely, they behave like pure propagating waves, which are *massless* objects! From the electric transport viewpoint, the maximum current density flowing in a carbon nanotube takes on 10 millions ampere per squared millimeter, namely, about 600 times niobium tin's, one of the best low-temperature superconducting alloys. It should be possible to build nanotubes based on boron and nitrogen. In contrast to carbon nanotubes, a BN nanotube could be extremely insulating and would pave the way for vast applications in terrestrial and space applications.

Going back to our solar sails, it is clear that nanotechnology-derived materials could supersede any concept of very thin all-metal or perforated sails (which nowadays would be highly desirable). It's a simple task to compute a (somewhat conservative) value for the nanotube-based sail loading: 0.02 g/m^2, or 5000 times better than the first solar-sail missions under study for the next few years. Achieving such a target would render the optical properties of the sail quite relaxable. In other words, the sail could be low in reflectance and full of wrinkles, though the sailcraft would still exhibit thrust accelerations tens of times that of the solar gravitational acceleration. So far, there is no sailcraft fast trajectory analyzed seriously for such levels of acceleration. Of course, nanotechnology materials and phenomena have to allow the designers to decrease in the same way the masses of the other main systems that the sailcraft is made up of. Even attitude control, power & communication, and environment–sailcraft interaction could reveal new results, completely beyond our current vision in designing both sailcraft and mission. Stay tuned!

The field of nanomaterials extends across the full range of traditional material classes, including ceramics, metals, and polymers. No previous materials technology has shown concurrent developments in both research and industry as do the areas of nanomaterials related to mechanical, electric, magnetic and optical components, quantum computing, biotechnology, and so on. There are very strong concerns about the capability of some molecules or particles to self-assemble at the nanoscale because they give rise to new substances with unusual properties. For such reasons, has *nanometrology* been recently introduced, namely, the ability to perform precise measurements at the nanoscale, an essential requirement for the correct and reliable development of nanotechnology in all its fields.

Using Ultimate Technologies

In science, the word *ultimate* is always risky to use, especially if one is trying long-term forecasting. It's very easy to wind up in science fiction. Therefore, here *ultimate* refers to post-nanotechnological science and technology, if any.

We noted above how atoms could be positioned with significant subnanometer precision. Were such capability achieved, one may talk of *picotechnology*, where, formally speaking, *pico* stands for 0.001 nm or 1 picometer (pm). By analogy, one may define picotechnology in the same way as nanotechnology, but in the range of 1 to 100 pm (the classical size of H-atoms is just about 100 pm = 0.1 nm = 1 Ångström (Å, after the Swedish physicist Anders Jonas Ångström), the unit preferred by atomic physicists). Thus, today, the term *picotechnology* refers to handling structures at the picoscale level, substantially with regard to high-precision positioning.

The next step would be femto-technology. Femtometer (fm) refers to one millionth of a billionth of a meter, also called the *fermi* in honor of the Italian physicist Enrico Fermi. The radii of atomic nuclei are expressed in femtometers. For instance, the radius of the carbon nucleus is approximately 2.75 fm, or about 0.00004 the related atomic "mean radius." Thus, femto-science is the current nuclear physics. In contrast, the claimed femto-technology might manipulate nuclei in a different way from what is already done in the laboratory by means of nuclear reactions. However, at the femto-scale, the so-called *strong* interaction (really hugely strong) dominates even in temporal terms. The characteristic times are many orders of magnitude lower than those achievable by the best atomic clock. The reader may think about femto-technology as an ingredient for a purely science-fiction book, including conjectured femto-weapons.

Nevertheless, there is an excellent exception that tomorrow's femto-technology may focus on: the *large-scale* production and storage of antimatter as the most energetic synthetic fuel. However, as explained in previous chapters, this is quite out of the capabilities of this century, unless real breakthroughs in physics occur.

As a consequence of the above considerations, currently there are no serious expectations coming from such technologies for advanced solar sailing.

Further Reading

Much of the historical material in this chapter has been reviewed or referenced in G.L. Matloff, *Deep-Space Probes*, 2nd ed., Springer-Praxis, Chichester, UK, (2005). A preliminary theory for the metallic perforated solar sail is also presented in that reference.

http://en.wikipedia.org/wiki/Nanotechnology.
http://en.wikibooks.org/wiki/Nanotechnology.
http://www.nanotechnologyfordummies.com/.
http://www.nano.gov/.
http://www.nano.org.uk/whatis.htm.
http://eoeml-web.gtri.gatech.edu/jready/main.shtml.
Nanotechnology, A Gentle Introduction to the Next Big Idea, 2002.pdf.
http://www.nsti.org/courses/ (for college students).
Springer Handbook of Nanotechnology, Springer-Verlag, Berlin, Heidelberg, New York, 2004 (for technical readers only).
The Handbook of Nanotechnology (Business, Policy and Intellectual Property Law), John Wiley & Sons, New York, 2005.

Progress to Date 13

At this point in its development, the solar sail can be characterized as fairly late in its theoretical phase and fairly early in its developmental phase. It is probably equivalent to the chemical rocket in 1930, the automobile in 1900, and the heavier-than-air aircraft in 1910.

Already, though, enough work has been performed for us to have some understanding of the basic possible configurations that might be considered for various sail applications. Also, the work of the last decade or so has indicated the potential roles of space agencies, private foundations and space societies, and private individuals in the historical and further implementation of space photon sailing.

Pioneering Designs

Figure 13.1 presents some suggested riggings, or configurations, for space sailcraft. These might be considered as celestial equivalents of terrestrial wind-sail riggings such as sloops and yawls

Starting from the top left of Figure 13.1 and moving clockwise, we first encounter the square rigged sail configuration. Here, solar-photon radiation pressure pushes against four sail segments supported by diagonal spars. The payload is mounted at the center, on either the sunward or anti-sunward sail face. It is not necessary to construct the spars and supporting structure from solid material—inflatable spars may be considered for many applications. Although square-rigged sails may be more difficult to deploy because they don't utilize centrifugal acceleration to push an unfurling sail from the center of the structure outward, the lack of spin may result in less dynamic problems such as vibrations and oscillations.

Next we come to the parachute sail, which carries its cable-supported payload on the sunward side of the sail structure. This is a more complex rigging to deploy and may therefore be utilized in space-manufactured rather than Earth-launched solar photon sails. Equipped with high-tensile

G. Vulpetti et al., *Solar Sails*, DOI: 10.1007/978-0-387-68500-7_13,

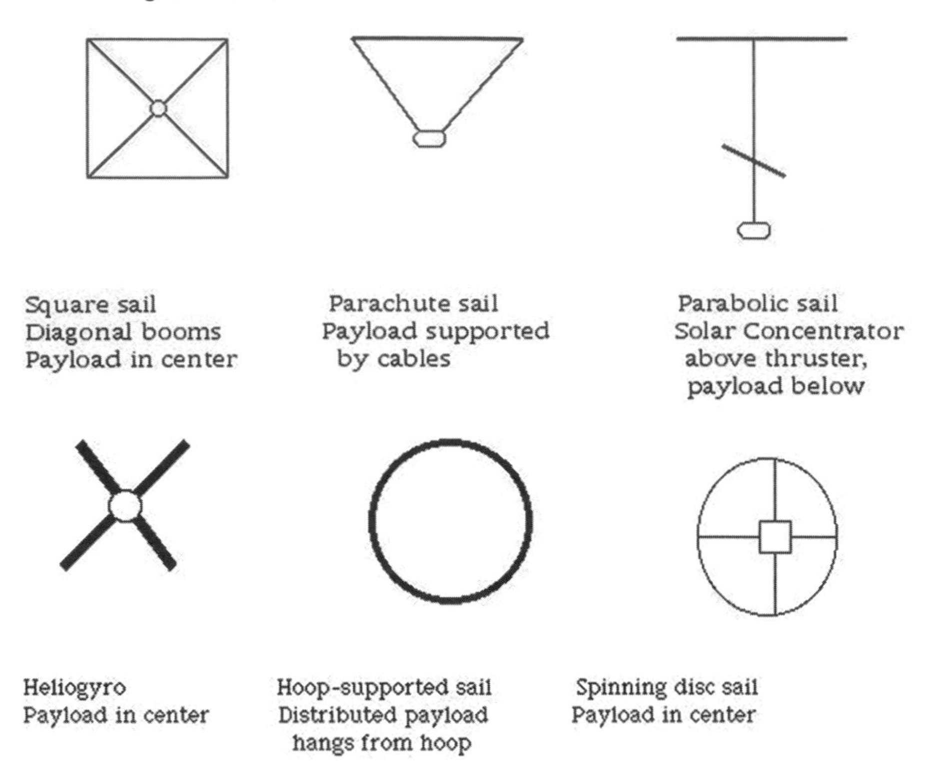

Figure 13.1. Various photon solar-sail configurations.

strength, low-density cables, parachute-type sails may be capable of higher accelerations than other arrangements.

The parabolic sail, or solar-photon thruster (SPT), is a two-sail variation on the parachute rigging. Here, a large sail or collector is always positioned normal to the Sun (or other photon source). The collector has a parabolic curvature (not shown in the figure) so that it can focus light on the smaller, movable thruster sail. A larger component of radiation-pressure–derived force can be tangent to the spacecraft's motion, allowing for this configuration's possible application in Earth-orbit raising. The SPT also has the potential to operate at a larger angle from the sunlight than other configurations. These advantages must be balanced against the added rigging mass and complexity.

Next is the spinning-disk sail. This rigging utilizes centrifugal acceleration as an aid in unfurling sail. The payload is mounted near the sail center.

A variation on the spinning-disk sail is the hoop sail. Here, the radial (possibly inflatable) struts are replaced by a hoop structure concentric to and containing the sail film. In this soap-bubble–like arrangement, the payload must be evenly distributed around the hoop structure, perhaps suspended from it.

The heliogyro sail rigging is inspired by the blades of a helicopter. After launch from Earth, the central core is slowly spun up and the blades are allowed to unfurl by centrifugal acceleration. Although sail deployment is relatively easy in this case, the blades must be long because of the comparatively small sail-film area-fill ratio. (There is simply not much of a sail for light to reflect from.) Payloads would likely be mounted near or at the sail's geometric center.

A final configuration is not shown in Figure 13.1. This is the hollow-body or inflatable sail. Here, a reflective film is mounted on the Sun-facing side of a balloon-like inflatable structure that is inflated in space using a low-density fill gas. The payload is near the center on the anti-Sunward-side of this "pillow." Although easy to deploy and mathematically model, hollow-body sails are more massive and more prone to micro-meteorite damage than other riggings.

Further investigative studies and operational applications will surely produce variations on the seven solar-photon-sail rigging arrangements considered here. But these seven will likely remain the basic approaches for the foreseeable future.

Although ultimate space-manufactured sail films may be very low mass monolayers, perhaps containing perforations smaller than a wavelength of light to further reduce mass, current candidate Earth-launched sail films are tri-layered. An aluminum layer about 100 nanometers in thickness faces the Sun and reflects 79 to 93 per cent of the incident sunlight (mainly depending on the surface roughness). Next comes a low-mass plastic substrate perhaps a few microns thick. On the anti-sunward side of this substrate is affixed an emissive layer (often chromium) that radiates the small fraction of sunlight absorbed by the aluminized face to the space environment.

In early sails, the plastic substrate is generally selected to be heat and vacuum tolerant and immune to the effects of solar ultraviolet (UV) irradiation. But there is a very innovative, mass-reducing suggestion to use instead a plastic substrate that sublimates rapidly when exposed to solar ultraviolet. Shortly after sail unfurlment, the plastic substrate would disappear leaving only a reflective-emissive bi-layer of very low mass.

This sublimation process, if controlled and unidirectional, could even add to sail thrust during the early phase of its journey. Called "desorption," this high-velocity sublimation of sail material is a subject of current research.

Because solar-photon sails (SPSs) are large-area devices that must accelerate for long periods of time in the space environment, a method of micrometeoroid protection has been developed. Similar to "ripstops" in terrestrial wind sails, a network of thin cables could be placed in the sail film. If a micrometeoroid impact were to destroy one small segment of sail defined by intersecting cables, other sail segments would still function.

Most early sail applications will involve low accelerations—probably in the vicinity of 0.0001 to 0.001 Earth surface gravities. But a 1996 computer finite-element-model study by Brice Cassenti and associates demonstrated that properly configured parachute, parabolic, and hollow-body sails are stable under accelerations as high as 2.5 Earth surface gravities.

Much work has been accomplished in sail design and much still remains to be done. But as the next sections indicate, government space agencies and private organizations have done much to remove this concept from the realm of science fiction and achieve progress toward the day when this innovative mode of in-space transportation will become operational.

The Role of Space Agencies

Much photon-sail research and development has been accomplished by national and transnational space agencies such as NASA and ESA. To better understand these contributions, it's a good idea to review the environment in which the space agencies operate.

The advantage of the space agencies over small-scale entrepreneurs is essentially one of scale. Since space agencies are governmental entities, they have the ability to plan long-term research and development efforts supported by tax revenues.

For example, much has been written in recent months about the success of privately funded suborbital space flights at a fraction of the cost of similar government-funded efforts. While these comparisons make a certain amount of sense, they entirely ignore the cost of the decades-long government-sponsored space infrastructure. Space Ship 1 would not have so readily won the X-Prize for repeated flights to heights in excess of 100 kilometers if Burt Rutan and associates had had to construct the Edwards Air Force Base and repeat the materials research leading to the technology used in their vehicle.

At least in democratic nations, however, this very advantage of space agencies to work with high annual operating budgets may work against rational space development. The space agencies must answer to the politicians, and the politicians must in turn justify expenditures to the electorate.

To get reelected, politicians must curry favor with the electorate. Sometimes high-profile stunts in space and huge projects economically supporting lots of workers with little practical output are favored over sounder approaches. The high-profile stunt results in favorable publicity and headlines; the "pork-barrel" project garners votes. To succeed, a rational space-development program must work with the politically inspired funding cycles.

With the exception of a few experimental efforts, all publicly funded space efforts utilize technologies mostly developed decades in the past. To allow new in-space propulsion technologies such as the SPS to mature to their flight application, NASA developed a step-by-step procedure called technological readiness, which works as follows: when a new space propulsion idea emerges from a theory and its basic physical principles are validated, it is assigned a technological readiness level (TRL) of 1. An example of an in-space propulsion concept now at TRL 1 is the proton-fusing interstellar ramjet. It may always remain at this level since its physics is well validated but its technology may never be defined. In some cases, such as the matter–antimatter rocket, the technological requirements can be defined, even if not achieved. Such propulsion concepts are at TRL 2.

As an in-space-propulsion concept matures, its TRL increases. Analytical or experimental proof-of-concept investigations are performed, followed by laboratory (breadboard) validation studies. Component and breadboard tests are then performed in a simulated space environment—a vacuum chamber—to achieve a TRL of 5. The next step is to successfully test a prototype of the in-space propulsion system under study in the simulated space environment. To achieve a TRL of 7, a prototype of the propulsion system must be successfully tested in space. The completed system is then qualified through demonstrations on Earth or in space. The highest level of TRL is 9, in which the propulsion system is operationally used in space missions. Examples of such "off-the-shelf" TRL- 9 propulsion systems include chemical rockets, solar-electric rockets, and gravity assists.

It might be argued that the TRL system is boring and bureaucratic—just the thing that a space agency might dream up to justify its own existence. But the beauty of the approach lies in its small, clearly documented incremental steps. A space-program manager can use TRL to compensate for the politically determined, highly variable nature of space-propulsion funding. Well-documented research can advance an in-space propulsion concept one or two TRLs during any funding cycle and then be used to efficiently pick up the research effort efficiently when large-scale funding resumes. In this way, it is not necessary to endlessly reinvent the wheel.

The ESA is used to applying the technology readiness procedure to new astronautical concepts through 9 levels as well.

World space agencies have done a great deal to advance the cause of the

SPS. In the late 1970s and early 1980s, NASA's Jet Propulsion Laboratory in Pasadena, California, analyzed the utility of the sail to perform a (canceled) 1986 rendezvous with Halley's Comet and propel a (canceled) extra-solar probe called TAU (thousand astronomical units). These paper studies led to the first tests of sail-like structures in space. In February 1993, a 20-meter-diameter thin-film reflector called Znamya was unfurled from a Progress supply craft docked to Russia's Mir space station. Znamya, designed to test the feasibility of reflecting sunlight to regions of the Russian arctic, was a modified heliogyro using centrifugal acceleration to unfurl.

In May 1997, an American space shuttle deployed a 14-meter-diameter inflatable antenna that tested the design of low-mass radiofrequency antennas and reflectors. Some of the concepts explored in this partially successful experiment are of relevance to inflatable, hollow-body solar sails.

The first test deployment of a true sail design in space came in the summer of 2004 when two small test sails were successfully unfurled from a suborbital Japanese sounding rocket. True to their country of origin, the sails were opened using the principles of origami, the Japanese art of paper folding! Capitalizing on this success, the Japanese space agency conducted an orbital solar-sail test in February 2006, when a test sail flew as a secondary payload aboard a rocket carrying the ASTRO-F (Akari) astronomical satellite. The sail unfurlment was a partial success.

Engineers affiliated with the NASA Marshall Space Flight Center in Huntsville, Alabama, have been raising the solar sail's TRL using a series of unfurlment tests of subscale sails in terrestrial vacuum chambers. During 2005, a 20-meter test sail was tested by NASA engineers in a terrestrial vacuum chamber (see Fig. 12.3 in Chapter 12). The pace of solar-sail development is quickening. And new players among government-sponsored space agencies can be expected to join the game. At present, we can safely conclude that the SPS has reached a TRL of 6 or 7 and that operational applications are not many years in the future.

But unlike the nuclear rocket, the SPS can be configured to any size. We might launch a micro-sail more properly called a solar kite that is not much larger than a living room rug with a payload of 1 or 2 kilograms. Our wealthier neighbor might at the same time be scaling the technology to propel an interplanetary ship with a sail diameter of 1 to 10 kilometers or even a larger interstellar craft.

With such a flexible in-space propulsion system, there is plenty of room for the small-scale inventor to make contributions, whether privately or governmentally funded. The next section considers the role of private initiatives in bringing the SPS to its current stage of flight readiness.

Private Initiatives

The early development of chemical rocketry was dominated by private inventors, such as Robert H. Goddard in the U.S., and national rocket societies in many countries. Private organizations and individuals continue to contribute to solar-sail progress.

A private individual or non-governmental organization has certain advantages and disadvantages when compared to government-sponsored space agencies. Since such groups or individuals are not beholden to taxpayers and politicians, they can tackle more visionary projects with a longer time to implementation or payoff. To implement these projects, however, private organizations must often engage in fund raising.

One contribution of private organizations has been raising public awareness of photon-sailing technology. Since 1982, three private groups—the Union pour la Promotion de la Propulsion Photonique (U3P) in France, the Solar Sail Union of Japan, and the World Space Foundation (WSF) in the U.S.—have collaborated to publicize the concept of a solar-sail race to the Moon.

Private organizations have also planned very nontraditional solar-sail propelled space missions. One American company (Team Encounter) has raised funds to launch human-hair samples on extrasolar trajectories, advertising that perhaps ethically advanced extraterrestrials intercepting the craft might feel compelled to clone the long-deceased human "crew" from the DNA in their hair samples. Very wealthy individuals might contribute to such a mission as a very-long-duration insurance policy!

But one of the greatest advances to photon-sail technology has resulted from the very serious work of the largest nongovernmental space organization of them all, the Planetary Society in Pasadena, California. Funded by member contributions and large donors including Ann Druyan (who is Carl Sagan's widow), the Planetary Society developed Cosmos 1, the first flight-ready spacecraft in which the photon sail would be the prime method of propulsion. To conserve funds, both the suborbital and orbital Cosmos 1 launches were conducted using a Russian booster of marginal reliability. Unfortunately, the reliability of this booster must now be classified as less than marginal since both launches failed and the sails plunged to Earth before they could be unfurled. The Planetary Society's directors hope to make another attempt with a more reliable booster. If Cosmos 1 does eventually achieve orbit, the pressure of sunlight will be used to alter the craft's orbit. One planned experiment is to beam microwaves to the orbiting craft using a radio telescope to experiment with collimated-energy-beam sailing. It would be nice if both solar and energy-beam sailing concepts can be validated on the same mission!

Temporary, small-scale organizations composed of visionary scientists and engineers have also contributed to the advancement of SPS technology and public awareness of this concept. During the 1990s, a group of researchers (including authors Vulpetti and Matloff) from several countries, met regularly in Italy to discuss the possibility of exploring nearby extrasolar space using sail-launched probes. It may be historically interesting to report how this team originated and worked. During the International Astronautical Congress, held in Graz, Austria, in October 1993, a group of seven solar-sail enthusiasts met to organize an in-depth study of solar sailing. After a lot of discussions, continued via mail for a couple of months, it was decided to set up a self-supporting study group. That meant that the group members would work during their free and creative time; nevertheless, some members would ask their companies to utilize some of the companies' facilities. Some companies said yes, and the group began working. The team chose the name Aurora Collaboration. (According to the ancient Greek mythology, Aurora was the younger, fair sister of Helios, the Sun god. Helios's elder sister Selene, the goddess of the Moon, was discarded for her paleness!) The active members of Aurora were author Gregory Matloff (NY University), Giancarlo Genta and his coworker Eugenio Brusa (Polytechnic University of Turin, Italy), Salvatore Scaglione (ENEA, Rome-Italy), Gabriele Mocci (Telespazio SpA, Rome, Italy), Marco Bernasconi (Oerlikon-Contraves, Zurich-Switzerland), Salvatore Santoli (International Nanobiological Testbed, Italian Branch, Rome, Italy), Claudio Maccone (Alenia-Spazio, Turin, Italy), and author Giovanni Vulpetti (Telespazio SpA, Rome, Italy). Vulpetti was appointed as the team coordinator. Aurora committed to the following objectives: (1) considering SPS propulsion for realistic extra-solar exploration; (2) investigating mission classes and related technological implications for significantly reducing the flight time, from departure to the target(s); (3) analyzing flight profiles; and (4) sizing sailcraft's main systems for a technology demonstration mission to be proposed to the space agencies. Aurora worked from January 1994 to December 2000. Some innovations have been developed and submitted to the attention of the space communities, including NASA and ESA. For instance, the NASA Interstellar Probe (ISP) concept (for which author Johnson served as the propulsion system manager) is an evolutionary development of Aurora. In turn, the current mission concept of the interstellar heliopause probe, in progress at ESA/ESTEC (Chapter 14), is similar to a smaller-scale version of NASA ISP.

The main results of Aurora, in chronological order, are as follows:

1. The fast solar sailing theory (in either classical or full relativistic

dynamics) and the related large computer code for optimizing unconventional trajectory classes

2. The bi-layer (Al-Cr) sail concept and the related preliminary experiments at ENEA for detaching plastic support in space, to have a clean all-metal sail
3. The concept of unfurling and keeping a circular sail via a small-diameter inflatable tube attached around the sail circumference; after sail deployment, the tube becomes rigid in the space environment and retains its shape without gas pressure
4. Sizing the onboard telecom system for communications from some hundreds of AU
5. The determination of the full behavior of aluminum's optical properties starting from experimental data
6. Optimization of trajectories to heliopause, near interstellar medium, and the solar gravitational lens

Aurora published 15 scientific papers, gave three presentations to European and Italian space authorities, and held a 1-day workshop at Rome University. Sometimes it is not necessary to resort to newspaper, radio or television advertising to foster genuine scientific advances. Serious, unheralded, and systematic work with pure vision and scientific objectives are still the basic ingredients for stimulating the appropriate institutions to transform good ideas into reality.

Further Reading

Two excellent sources considering in greater depth the material covered in this chapter are Jerome L. Wright *Space Sailing*, Gordon and Breach, 1992, and Colin McInnes *Solar Sailing*, Springer-Praxis, Chichester, UK, 1999. More information on various sail configurations can be found in the appendix of Gregory L. Matloff *Deep-Space Probes*, 2nd ed., Springer-Praxis, Chichester, UK, 2005.

Future Plans 14

> We have lingered for too long on the shores of the cosmic ocean; it's time to set sail for the stars.
>
> —Carl Sagan, Cosmos television series, 1980

The Next 25 Years

While our technology will not yet let us "set sail for the stars," it will let us take the first steps—though, from our perspective, at a snail's pace. Lack of political will, tight funding, and competing scientific objectives may make it likely that solar-sail propulsion will not be used significantly for at least another quarter of a century. However, one could be optimistic also in light of the tremendous technical accomplishments in the field over these last few years, and speculate that humanity will be sending sails throughout the solar system within a decade. But others could say that this optimism would ignore the sad reality of the last 40 years of space exploration that has seen the introduction of only a few new propulsion technologies, each taking nearly a decade to go from demonstration in space to their use on a second mission. Finally, one should note that many space-faring countries have near-term serious solar-sail programs. Let us look at the world panorama at the time of this writing.

United States

A solar-sail demonstration mission was competing within NASA for flight validation as part of the New Millennium Program (NMP). The solar sail was one of five technologies competing for flight in the year 2010 in what the program calls Space Technology 9 (ST-9). If it had been successful, the sail demonstration would have been allotted approximately $90 million to perform an orbital test of solar sailing, using the inflatable boom technology of L'Garde, Inc. (See Chapter 12 for more information about L'Garde and its inflatable boom technology.) Regrettably, ST-9 was cancelled. Would this

G. Vulpetti et al., *Solar Sails*, DOI: 10.1007/978-0-387-68500-7_14,

mission demonstrate everything needed to next implement one of the missions listed in Chapter 9? Probably not; the funding was simply not sufficient to flight-qualify each and every subsystem at the level that would allow a science mission to simply copy the design and use it. This means that any follow-on mission would have to commit some of its resources (money and development schedule) to addressing the unanswered questions remaining from the validation flight and complete the sail design process. Why can't an ST-9 solar-sail mission accomplish more? The simple answer is lack of funding. The NMP is allotted only a finite amount of funds per flight, and it must accomplish what it can within these constraints. Is the validation flight worth the cost? The answer must be yes if it successfully demonstrates the design, development, manufacture, launch, space deployment, navigation, and model validation necessary to scale the design to larger sails in the future.

The Planetary Society experienced a significant setback in its effort to fly a solar sail when its Russian rocket malfunctioned, resulting in the sail's never reaching space. But, to quote the Planetary Society's Web site, "this does not spell the end" of their effort to build and fly a sail. The Planetary Society is populated by visionaries and led by an experienced space advocate and former NASA Jet Propulsion Laboratory (JPL) project manager, Dr. Louis Friedman. The membership base raised the funds for the Cosmos 1 solar sail, and is hard at work doing the same for another launch attempt. Even if Cosmos 1 were completely successful, it would only be a demonstration of the technology (though an important and significant demonstration), and not a prototype for future missions to use again.

Japan

The Japanese are moving forward in solar-sail development, building upon their successful subsystem space tests in 2004 and again in 2006. It is possible that Japan will fly a full-scale solar sail within the next 5 years. If it does fly such a sail and the mission is successful, then it is possible we might see a new sail mission once or maybe even twice in the following decade.

Russia

The Russian space mirror program is in hiatus and with it the efforts to fly solar sails—with one notable exception. Providing both funds and technical know-how, the Planetary Society went to Russia to build the Cosmos 1 solar sail. Using this expertise, combined with the successful Znamya mission, the Russians have demonstrated that they know how to fabricate a solar sail for space. When, if ever, will they fly one themselves?

Europe

The ESA, fostered mainly by industries and field professionals from Italy, Germany, United Kingdom, and France, has a number of plans for solar sailing. In May 2006, ESA launched an invitation to tender proposals for phase A of the Geosail mission, which was described in Chapter 9. The ESA would allot 150 million euros (about $190 million) for phases B, C, and D of this program. Geosail is not only a technology demonstration mission, but also a mission of very high scientific concern seeking knowledge of Earth's magnetosphere and new phenomena in magnetized plasmas. The Geosail operational orbit would be an elliptical orbit 11 × 23 Earth radii (namely, well beyond the geostationary circular orbit, the radius of which is 6.61 Earth radii), completely controlled by solar sail for 3 to 5 years. Geosail is a full mission with a number of technological and scientific unknowns, and this renders it still more intriguing. Almost all the technologies that are proven in space by this mission will likely be considered for any subsequent missions.

As a point of fact, there are two other main solar-sailing studies in progress at ESA: (1) the solar polar orbiter, and (2) the interstellar heliopause probe (IHP). Such missions are very challenging. Mission A-1 would utilize a solar sail to lessen its initial orbit to less than 0.5 AU before raising its inclination. Then, changing the sail orientation, the latitudes of the solar poles can be achieved. The sail will be jettisoned after the operational circular orbit about the Sun is achieved. Such a mission is one of several examples of "missions impossible" for rockets (See Chapter 3). Mission B-2 is similar to NASA's interstellar probe (ISP) with 200 AU to be reached after 25 years from launch. However, at present, the IHP sail is not envisaged to be as lightweight as its American counterpart. As a consequence, ESA-IHP would be required to fly by the Sun two or more times to get the energy needed to achieve the desired flight time; in any case, the cruise speed is significantly lower than that of the NASA interstellar probe.

Europe's current efforts may develop considerably more than what is planned today. Much depends on the Geosail mission.

What about nongovernmental groups other than the Planetary Society? There have certainly been quite a few of them appearing in the media these last several years, promising everything from demonstrating sails in space for advertising purposes to offering to take your DNA to the stars. One of the most prominent, even receiving thousands of dollars in contracts from the U.S. National Oceanic and Atmospheric Administration was Team Encounter. Team Encounter planned to use the solar-sail technology of L'Garde to send a 3-kilogram payload on a solar system escape trajectory. The payload was to include DNA from selected donors, as well as messages from of people all over the world. To be launched as a secondary payload on an Ariane 5 rocket, the craft would deploy a 4900 m^2 sail and head toward the edge of the solar system. The Team Encounter mission was originally slated for launch in the early part of this decade, but its status is now unknown.

With the advances in solar-sail technology—in the United States by NASA, the Planetary Society, and others; in Europe by DLR and Russia; and in Japan—we might be on the edge of seeing their widespread use for a variety of missions. Certainly the technology is here. We've built large sails and tested them in simulated space environments, we've flown the materials in space, and some have even spun up and deployed sails in low Earth orbit. Unfortunately, these are all subscale, low-cost technology demonstrators that cost significantly less than would their larger, more capable cousins needed for true exploration of deep space. When might we see solar sails in widespread use? Perhaps we can look at the past and make an educated guess about their future. The experience of gridded-ion solar electric propulsion might serve as a good example of a new propulsion technology that followed a path similar to what we might expect for solar sails.

Why did it take 5 to 9 years for ion propulsion to move from space validation to mission use? There are many reasons; most of them make good logical sense and most will apply to solar sails. For advocates of new technologies, they feed a sense of growing frustration.

Reason 1: Timescale

A typical robotic space mission requires 3 to 4 years of development, from selection to flight, and more years after that for analysis of the data. Given that these missions are relatively expensive, costing many millions of dollars each, not many are selected and flown. And not many scientists want to risk 10 years of their career on an unproven space propulsion technology. Since ion propulsion technology was not "proven" until it flew successfully in space, no one proposed its use until after completion of the mission. Even if the next mission were announced on the day DS-1 launched (which did not happen), it would have been a minimum of 3 years before that next mission would have been ready to fly. In fact, it was unlikely that the next mission would be proposed until after the DS-1 mission was complete and the data regarding the propulsion system was published, adding another 2 years to the wait. The same "fly it and see" attitude will likely be present for solar sail propulsion as well.

Reason 2: Science

Within NASA and other governmental space agencies, the mission selection process is not based on "cool technology." Robotic missions are selected for

First proposed by Dr. Robert Goddard in the early 1900s, the idea of propelling a spacecraft with electrically charged atoms, called ions, instead of chemical rocket exhaust, became a reality as the primary propulsion system for a deep-space spacecraft with the flight of the Deep Space 1 (DS-1) mission in 1998. Using electrical energy generated onboard the spacecraft to power its ion drive, the DS-1 spacecraft demonstrated the use of a gridded-ion propulsion system for the first time. Noted for its very high efficiency, more efficient than a chemical rocket by at least 10:1, ion propulsion is ideally suited for the exploration of deep space. Incapable of lofting a spacecraft into space from the surface of Earth due to its very low thrust, ion propulsion is hard to beat for some (rocket-based) missions in the outer solar system due to its high exhaust speed and resultant ability to deliver twice as much payload at a destination when compared to its chemical propulsion counterpart.

With these significant mission benefits known, it nonetheless took about 50 years for this technology to go from the laboratory to flight. Members of the Werner von Braun rocket team at the Marshall Space Flight Center in Huntsville, Alabama, conducted early ion propulsion experiments in the 1960s. With that center's emphasis on sending humans to the moon as part of the Apollo Project, the work on electric propulsion was sent to NASA's Lewis Research Center (now the Glenn Research Center) in Cleveland, Ohio. Scientists and engineers at Glenn worked diligently on the technology until it was flown on the first mission of the New Millennium Program, DS-1, in 1998. Following its successful flight, expectant engineers anticipated mission after mission to baseline the technology and for a new age of efficient, deep space exploration to begin. Instead, no NASA mission selected the technology for flight until 2001. Dawn, a mission to visit the asteroids Vesta and Ceres, would not be possible if it were not for this highly efficient ion propulsion system. Dawn was launched in 2007—nine years after the DS-1 mission first demonstrated ion propulsion as a viable in-space propulsion system.

Not waiting for NASA, the Japanese launched the Hayabusa mission in 2003. The gridded-ion propulsion system on Hayabusa was intended to allow it to land on the surface asteroid Itokawa (Fig. 14.1) and return a small sample from it back to the earth. Due to an in-flight fuel leak, the mission was compromised and the spacecraft's ability to make the return voyage to Earth is in doubt. The ion propulsion system on Hayabusa performed well. This mission flew 5 years after the completion of the DS-1 mission.

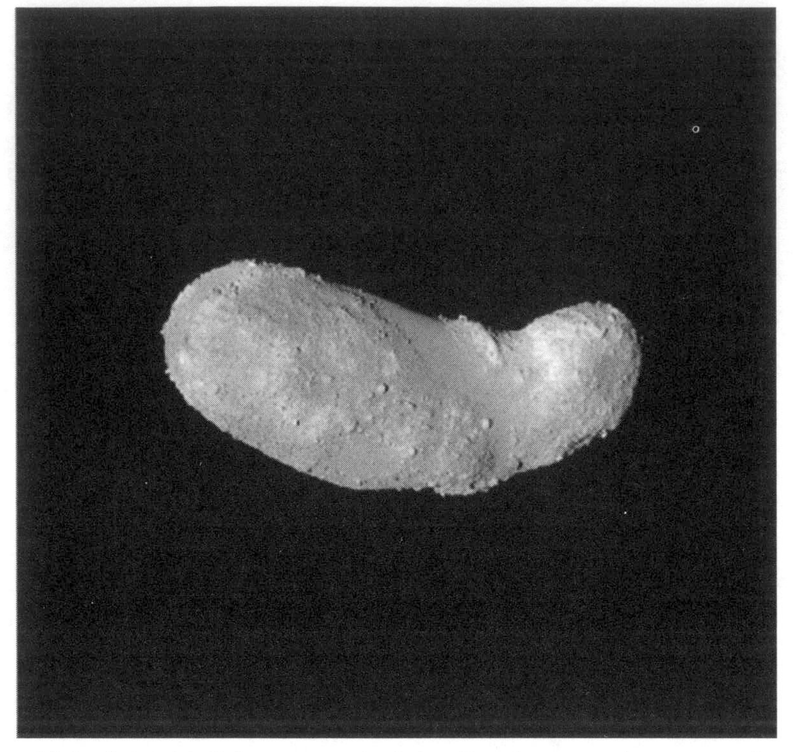

Figure 14.1. Asteroid Itokawa, reached by the Japanese ion-powered probe Hayabusa in September 2005. The probe was launched in May 2003. The Hayabusa's return to Earth is expected for June 2010. (Courtesy of JAXA)

the most part on the basis of the science they propose to return. A mission to use a new technology, be it ion propulsion or solar sails, will have to compete for funding with missions that use neither and whose science will be at least as compelling as that using the new technology. It might be years before another mission with the "right" science and that uses the new technology is selected for flight. Such was the case within NASA and the selection of the Dawn mission.

Reason 3: Risk

The number of deep space missions flown with ion propulsion at the time Dawn was competing for funding stood at exactly one. Chemical propulsion has been used on hundreds of missions, many of which went beyond Earth orbit. Chemical propulsion systems are simply better understood and their risks are better known than those related to solar electric propulsion systems. How can we know if the DS-1 succeeded because of a good design and engineering versus simple luck? With only one data point, we cannot know. For this reason, a new propulsion system with limited (even successful) flight history will be considered higher risk than others when vying for flight status. In an evaluation process, such a mission will have to overcome the problem of being considered "higher risk" even if it promotes more compelling science.

The above three reasons should not be taken systematically as the primary criteria of choice; otherwise, if one relies always on the past achievements, no progress would be possible in any area of the human activity.

The Next 50 Years

Making the most optimistic assumptions, that solar sails are flown in space before 2010 and that the second mission flies within 6 years of the first, we should expect no more than two or three more solar sails to fly before the first quarter of the century passes. These first sails will be built from today's technologies and perform precursor missions, either geocentric or near-Sun.

In the years to follow, we will see the introduction of the next generation of solar sails, which will be bigger, lighter, and more capable than those of the first generation. It is in this time frame that we might see a mission to the edge of the solar system such as the proposed interstellar probe. (The NASA interstellar probe mission is described in Chapter 9.) It seems to be unlikely that more ambitious missions to the solar gravitational lens or to the Oort cloud will occur until beyond even this time frame.

At the end of the next half-century, we may see the advent of space settlements and the ability to fabricate large, gossamer structures using the resources provided in space—from materials mined on the moon or asteroids. If so, then sails that perform much better than those made in Earth's gravity field might be fabricated, allowing more payload or faster trip times to multiple destinations in the solar system and beyond.

As emphasized in Chapter 12, the classical vision of solar sailing might be enlarged or superseded, in some cases, by different concepts like the nanosailcraft swarm. Fifty years are sufficient to surprise us!

The Next 100 Years

Foreseeing things over so long a time is always very difficult, not only because extraneous factors may come into play, but also since a new understanding of physics could revolutionize many human activities, including spaceflight.

Considering the current physics, as our ability to manufacture ever-larger and lighter sails develops, we will see their use crossover from purely robotic missions to those that support the expanding human presence in the solar system. Sunlight-propelled cargo ships, a stream of them carrying cargo between Earth and Mars, might crowd the space between us and our settlements on the red planet.

Thrown into the mix will be an asteroid or comet, slowly diverted from its orbital path by a solar sail to either avert global catastrophe or provide raw materials to our burgeoning interplanetary civilization.

We might even see the construction of a massive laser or microwave power beaming station with the goal of using the beamed energy to send a small probe deep into interstellar space or, perhaps, to a nearby star.

Space Sailing: Some Technical Aspects

The first three parts of this book have described solar sailing and sailcraft, as well as the design problems, the unknowns, new perspectives, and the expectations that a revolution in space propulsion and vehicle design may be one of the main keys for an extended exploration and utilization of space within a few decades. The level of topic presentation was kept as simple as possible to provide nontechnical readers with basic information in every major area of space sailing without becoming involved in the underlying mathematical constructs.

However, like any area of science and technology, a deeper knowledge entails higher concepts and a more appropriate language. The universal language of science is mathematics. In general, different though interconnected mathematical disciplines are used for addressing specific topics. A problem may be dealt with through many steps, each step obeying the underlying set of different and progressive assumptions of the current model.

Part IV is intended for more technical readers, in particular for undergraduate students in physics, engineering, and mathematics. However, the math has been kept to a simple level. To read the following chapters requires a modest background in physics, and elementary calculus is also advisable. The following chapters should be viewed as a short introduction for students interested in the dynamics of solar sailing as part of their future professional activity. In such a context, all the topics addressed in this book could also aid the reader to get a sufficiently general view of the problems related to both solar-sail spaceflight and the next steps of the endless human adventure in space.

Space Sources of Light

15

In Chapter 5, we addressed the problems of light and its amazing nature. We discussed the twofold nature of light: wavc and particle. As this book regards solar sailing as a non-rocket photon-driven propulsion mode, we shall primarily focus on the properties of solar light and, secondarily, on the light from planets.

The energy of a photon is directly proportional to its frequency or, equivalently, inversely proportional to its wavelength. Frequency and wavelength refer to the oscillations of electric and magnetic fields traveling in a vacuum or inside matter. Wavelength determines the way both fields interact with objects met along their propagation path. As a consequence, one can divide the electromagnetic spectrum into regions or bands (with different names). Bands are divided into sub-bands; historically, their nomenclature changed according to the progressive knowledge of their features. Figure 15.1 shows the main regions of the spectrum in terms of wavelength expressed in nanometers (nm), microns (μm), or centimeters (cm), according to the band. Also, some of the sub-band names have been reported. Note how small the visible band is (0.4—0.7 μm) compared to the other regions. It may seem incredible that a typical TV wave transports energy 10^{-12} times lower than a photon of 0.1 nm wavelength, namely in the x-ray region. Someone might objcct that the spectrum regions/subregions are somewhat arbitrary. This is only partially true. Although discussing the related criteria is beyond the scope of this chapter, wc mention an example: the gamma-ray region begins at 511 keV, which corresponds to the energy of the rest-mass of the free electron. The wavelength of any photon carrying this energy in vacuum is equal to $\lambda = hc/E = 1239.84191\ \mathrm{nm}/E[eV] = 0.002426$ nm. (In practice, though, neither wavelength nor frequency is suitable for featuring photons beyond the subregion of the hard x-rays, but energy is appropriate).

Considering the importance of the concepts regarding energy emission from a source of light and the energy received by a surface, we introduce the following definitions:

G. Vulpetti et al., *Solar Sails*, DOI: 10.1007/978-0-387-68500-7_15,

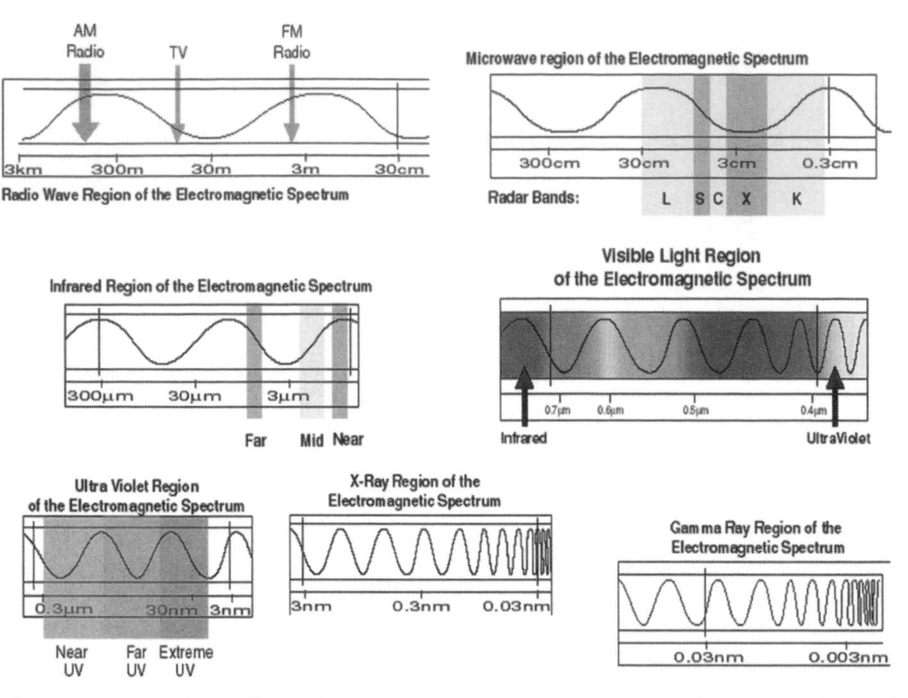

Figure 15.1. Regions of the electromagnetic radiation spectrum. (Courtesy of NASA)

1. **Source of light**: Photon radiation can be emitted by any body endowed with a temperature, not necessarily uniform or stationary. A source can be the surface of proper photon sources (like stars, lamps, usual hot bodies, gas, plasma, etc.) or any surface reflecting/scattering a fraction of the received light. When the source does not appear as point-like, the emitting surface can be partitioned in elementary or infinitesimal surfaces, each of which is endowed generally with its own radiation characteristics. Given an elementary oriented surface **dA**, emitting or receiving energy, and a direction **d** of radiation emission or incidence, the angle of emission or incidence between **d** and **n**, the positive normal to **dA**, is denoted by θ (also called the zenithal angle). Thus, the projected or orthogonal-to-**d** area is equal to $dA_n = \cos\theta \, dA$, dA being the magnitude of **dA**.
2. **Radiant power** or **Radiant flux** (Φ): the total power, expressed in watts, emitted by a radiation source. It does not contain any other source-related information. Note that, differently from other definitions below, Φ is the basic quantity by which one can build the other ones. In particular, it cannot be defined *after* Φ_λ.
3. **Spectral radiant power** (Φ_λ): the power emitted per unit wavelength. $\Phi_\lambda = d\Phi/d\lambda$, [W/µm].

4. **Radiant intensity** (F): measured in W/sr, it is the power per unit solid angle emitted by a radiation source. $F = d\Phi/d\omega$. It should not be confused with radiance.
5. **Radiant emittance** or **radiant exitance** (M): the power emitted per source's unit surface [W/m^2]. $M = d\Phi/dA$. This power is assumed to radiate into the hemisphere that contains **n**.
6. **Spectral radiant intensity** (F_λ): the radiation power emitted per unit solid angle and unit wavelength. $F_\lambda = dF/d\lambda = d^2\Phi/d\omega d\lambda$, usually measured in [W/sr nm or W/sr μm].
7. **Spectral radiance** (L_λ): power emitted by a radiation source per unit wavelength, unit solid angle, and unit *projected* area. Spectral radiance is expressed in [W/(m^2 sr nm)]. $L_\lambda = d^3\Phi/dA_n\, d\omega\, d\lambda$. Depending on the problem at hand, either photon frequency or energy can be used instead of wavelength.
8. **Radiance** (L): the spectral radiance integrated over a range or band of wavelengths. In SI units, radiance is expressed in [W/(sr m^2)]. Note that L could first be defined as $L = d^2\Phi/dA_n\, d\omega$, and then the spectral radiance would follow as $L_\lambda = dL/d\lambda$ at givenλ. Radiance should not be confused with the radiant intensity. For a Lambertian (flat) surface, L is independent of the viewing direction, by definition; as a result, it comes out that $M = \pi L$.
9. **Spectral irradiance** (I_λ): power per unit wavelength incident on or crossing a unit surface. It is expressed in [W/m^2 nm] or [W/m^2 μm]. $I_\lambda = d^2\Psi/dA\, d\lambda$, where power Ψcomes from all directions in either hemisphere based on **dA**.
10. **Irradiance** (I): the spectral irradiance integrated over a range or band of wavelengths (or the broadband irradiance). When the band is the full electromagnetic spectrum, one gets the total irradiance. Again, one could first define the irradiance and, then, the spectral irradiance as $I_\lambda = dI/d\lambda$ at given λ. Irradiance is measured in [W/m^2].

The concepts expressed in the above definitions are fully compliant with the regulations of the *Commission Internationale de l'Eclairage* (CIE, International Commission on Illumination) for radiometry and photometry. (The latter is radiometry restricted to the visible band, but connected to the spectral sensitivity of the human eye.) In particular, definitions 1 to 8 address the sources of light, whereas definitions 9 and 10 address radiation received by a surface (even an ideal one). However, a few scientific communities may have adopted different terminology and meanings. For instance, meteorologists call *flux* (as shorthand for *flux density*) the rate of radiant energy passing through a given flat surface, expressed in [W/m^2].

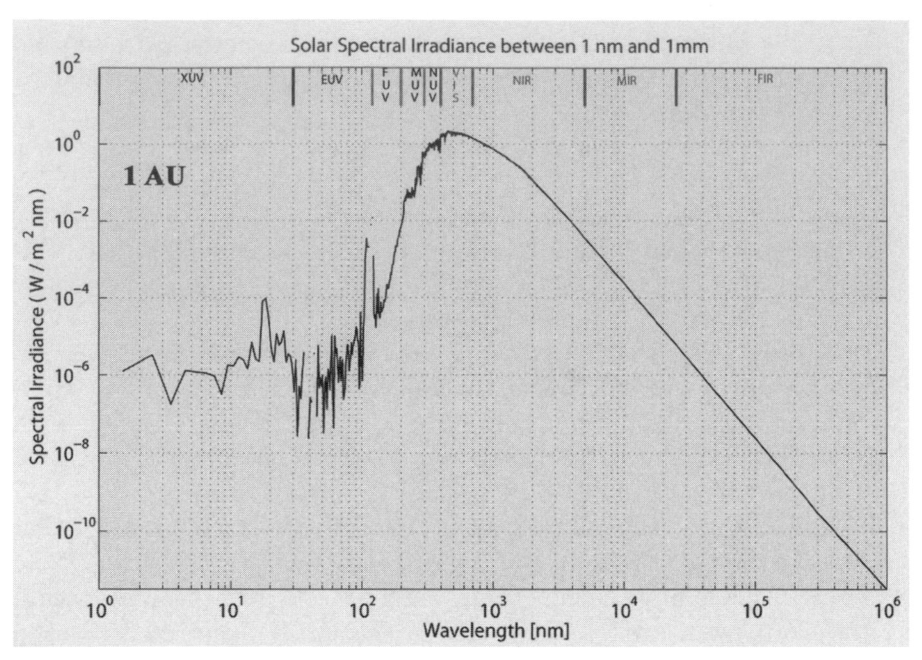

Figure 15.2. Sun's spectral irradiance [W/m^2 nm], at 1 astronomical unit (AU), over six orders of magnitude in radiation wavelength. Regions have been labeled.

Consequently, they define the incident flux per unit solid angle [W/m^2 sr] as the radiant intensity impinging on a given area.

We use the international conventions and units here. The concepts of spectral and total radiance and irradiance will suffice for our purposes in this chapter.

Like any other source of natural electromagnetic radiation, the Sun does not emit light uniformly over wavelength. If one considers a unit surface placed at 1 astronomical unit and orthogonally to the sunlight propagation direction, it is possible to measure the energy that crosses such area per unit time, totally or as a function of the wavelength. In the former case, one gets the *total solar irradiance* (TSI), whereas in the latter case one obtains the *solar spectral irradiance* (SSI). Figure 15.2 shows SSI, in units of W/(m^2 nm), from 1.5 to 10^6 nm. The subregions indicated in the upper part of the figure are detailed in Table 15.1. In other books, you can find some differences in the reported ranges. For instance, the 100 to 400-nm range can be divided into ultraviolet (UV)-C, UV-B, and UV-A, also according to the health effects on the human body. The ranges reported in Table 15.1 are compliant with International Standards Organization (ISO) initiative 2002.

From Figure 15.2 and Table 15.1, one can note that (1) the visible band encompasses most of SSI; (2) there are strong variations in the overall UV region; (3) SSI decreases monotonically in the full infrared (IR) band; and (4) almost 92 percent of TSI resides in the [0.4–25] micron range, whereas the UV range captures 8 percent of it. Such figures are important in choosing the reflective layer of the sail.

Table 15.1. Wavelength ranges of some subregions of the electromagnetic spectrum

Fraction of the total Irradiance	Subregion	Min λ [nm]	Max λ [nm]	Notes[a]
	XUV (soft x-rays)	1	30	Ionizes atoms and molecules; absorbed in the upper atmosphere
	EUV (extreme ultraviolet)	30	120	Ionizes nitrogen and oxygen molecules; absorbed above ~90 km
	FUV (far ultraviolet)	120	200	Dissociates oxygen molecules; absorbed above ~50 km
0.08	MUV (middle ultraviolet)	200	300	Dissociates oxygen and ozone molecules; absorbed between ~30 and ~60 km
	NUV (near ultraviolet)	300	400	Can reach the ground
0.39	VIS (visible)	400	700	In practice, passes unabsorbed through the full atmosphere
	NIR (near infrared)	700	4000	Partially absorbed by water vapor
0.529	MIR (middle infrared) or thermal infrared	4,000	25,000 or 50,000	Absorbed and re-emitted by ozone molecules, CO_2, water vapor and the other gases present in low atmosphere
$<5\times10^{-5}$	FIR (far infrared)	25,000 50,000	1,000,000	Fully absorbed by water vapor

[a] Notes refer to Earth's atmosphere.

The values of spectral energy in Table 15.1 change over a solar cycle. Such variations are important for Earth's atmosphere, especially the UV range,

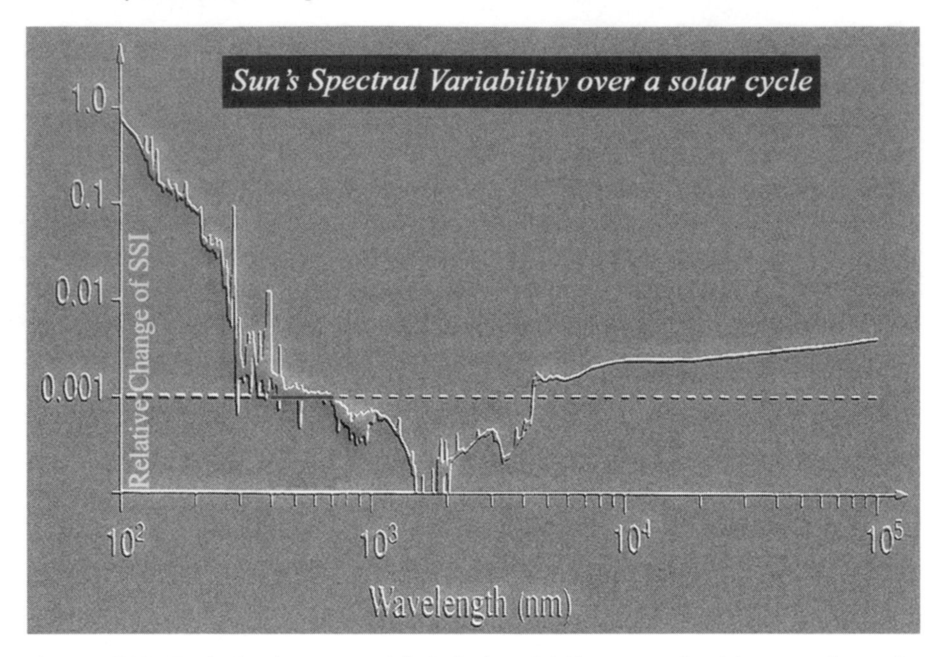

Figure 15.3. Typical solar spectral (relative) variability, over the 11-year solar cycle, from the upper extreme ultraviolet region to the lower far infrared region. The dashed horizontal line represents the visible radiation change (approximately).

inasmuch as its energy content represents the input of energy to the upper atmosphere: UV photons deposit their energy in the atmosphere layers known as the stratosphere, mesosphere, and thermosphere (altitude increases from the first one). In particular, they make and maintain the ionosphere. Figure 15.3 shows the solar spectral variability over one of the latest solar cycles (numbers 21 to 23), which have been observing since November 1978 by means of high-sensitive instruments onboard satellites orbiting Earth or around the L1 point of the Sun–Earth system. The relative change of SSI is plotted versus wavelength from 100 nm to 100 μm. The horizontal red segment, which fits the band from 400 to 700 nm, represents the mean change (0.1 percent) of the visible light. It is striking how much the overall ultraviolet band can change during a solar cycle. This contribution causes strong variations in the upper atmosphere layers.

From a solar-sail propulsion viewpoint, the above discussion entails that the sail's reflective film has to mainly reflect the visible and infrared bands; also, it has to be resistant to ultraviolet photons for decreasing optical degradation. In Parts I to III of this book, statements like this one are now justified by modern measurement campaigns.

Let us gain additional information about solar light. Figure 15.4 zooms in

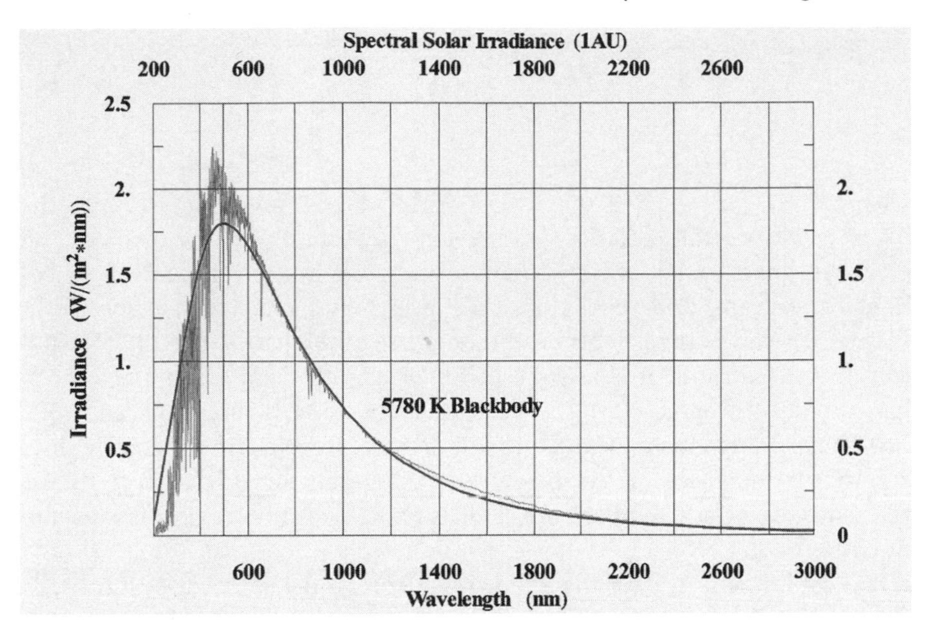

Figure 15.4. Solar spectral irradiance (1 AU) from 200 to 3000 nm. The smooth curve is the blackbody distribution that produces the same total irradiance (approximately 1370 W/m^2) over the full solar spectrum.

to the central part of Figure 15.1, namely, from 200 to 3000 nm. The orange line represents the observed SSI, whereas the magenta smooth line denotes the spectral radiance, integrated over the solid angle the Sun subtends at the 1 AU (or 6.80×10^{-5} sr), of a blackbody at 5780 K. Such plots give us some important information: (1) the Sun behaves as a blackbody of high temperature, which refers roughly to the photosphere one observes as a whole (as a point of fact, even slightly below or above or locally on the photosphere, temperatures of active solar zones can be much different); (2) the infrared band follows the blackbody distribution pretty well; (3) the visible and ultraviolet bands show nonnegligible deviations from the blackbody, especially from the ideal maximum, that is, ~ 1.8 W/(m^2 nm) at 500 nm.

Blackbody distribution features a one-to-one relationship between spectral radiance and temperature. In other words, if you know the radiance L_0 of an emitting body (regardless of its real properties) at some wavelength λ_0, there is one temperature T_b distribution passing through the point (L_0, λ_0). Thus, a general emitter can be characterized by a distribution of blackbody temperatures corresponding to its real radiances values. This is the concept of *brightness temperature* of a real emitting body. In formal terms:

$$T_b = B^{-1}(L_0, \lambda_0) = \frac{hc/k}{\lambda_0 \ln\left(1 + 2\frac{hc^2}{L_0 \lambda_0^5}\right)} \tag{15.1}$$

where $B^{-1}(L_0, \lambda_0)$ denotes the inverse of the Planck function describing the blackbody spectral radiance (which can be found in any textbook on electromagnetic radiation). In Equation 15.1, c, h, and k denote the speed of light in vacuum, and the Planck and Boltzmann constants, respectively. Figure 15.5 shows the brightness temperature of the Sun in the 200 to 3000-nm range (the same of Fig. 15.4). In this range, the Sun looks like a set of blackbodies from 4000 to 6,450 K. Note, comparing Figures 15.4 and 15.5, how small deviations in radiance at longer wavelengths translate into significant increases of the brightness temperature with respect to the reference blackbody temperature. This is expressed in the rightmost part of Equation 15.1.

Now it's time to answer the following question: If a solar sail is R (usually astronomical units) away from the Sun, what is the total solar irradiance on it? Denoting such irradiance by $I(R, \mathbf{p})$, one can write

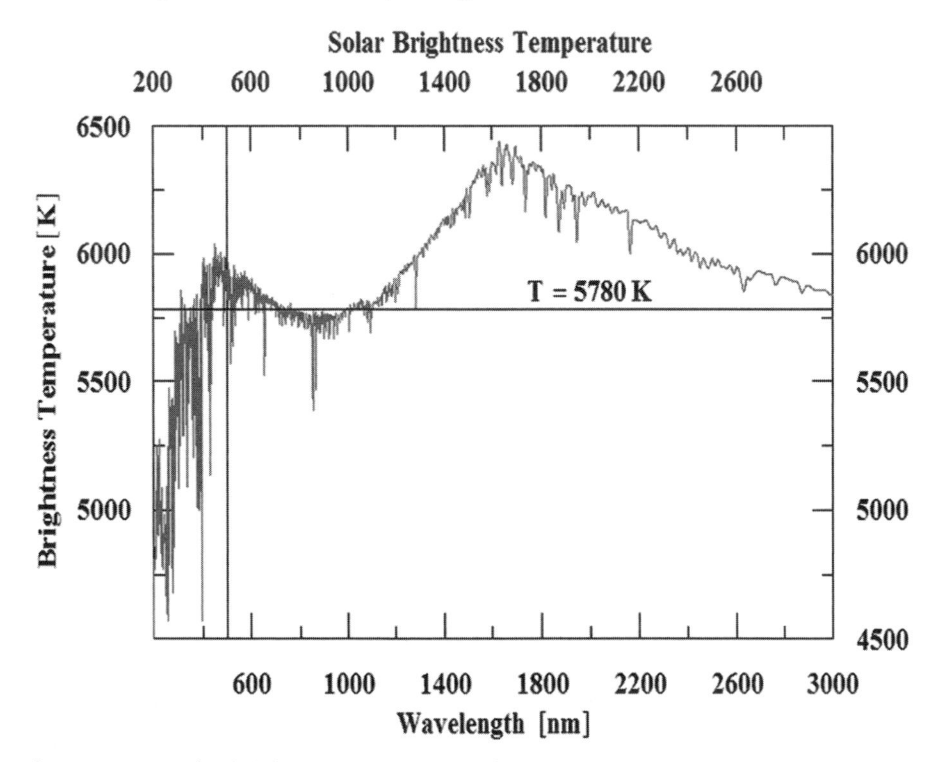

Figure 15.5. Solar brightness temperature from 200 to 3000 nm.

$$I(R, \mathbf{p}) = I_{1\,\mathrm{AU}}\, f(R, \mathbf{p}) \qquad (15.2)$$

where the first term on the right-hand side is the standard TSI, whereas the second term represents the scaling function. The set of parameters the scaling factor depends on has been denoted by **p**. Figure 15.6 shows the behavior of TSI in the solar cycles 21 to 23. The colored lines plot the daily averaged values coming from the radiometers on board satellites. The black line comes from data smoothing. Such a figure is the composite TSI time series, made by PMOD, which unifies measurements from different radiometers and histories (i.e., including degradation), and is adjusted to 1 AU. TSI, also named the solar constant, is not a constant; this quality jump—fully related to the space era—began on November 16, 1978, by means of the Hickey-Frieden (HF) radiometer on the satellite Nimbus-7. Subsequently, other high-precision satellite radiometers have measured the total solar irradiance every 2 to 3 minutes; as of November 2006, the radiometers Virgo on the spacecraft Soho have been continuing to monitor the Sun. In the next 2 years, the French microsatellite Picard and NASA's large Solar Dynamics Observatory (SDO, under the Living-with-a-Star

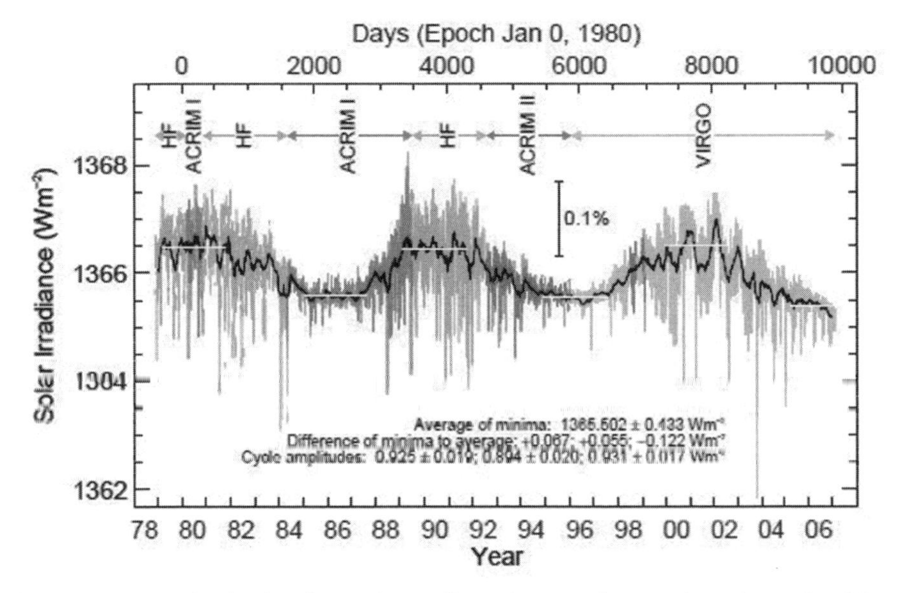

Figure 15.6. Total solar irradiance (TSI, adjusted at 1 AU), over the solar cycles 21 to 23. The colored lines are the composite daily averaged values from many satellite observations. The black curve represents data smoothing. Note that the cycle amplitudes are lower than 1 W/m^2. Such space-borne data of TSI have been showing that the solar constant is not constant. (Courtesy of PMOD, World Radiation Center)

program) will be launched for new campaigns of solar radiation measurements for better understanding the complex solar activity cycle. Forecasting TSI over the next several years is a difficult task indeed. However, reliable data, such as those ones plotted in Figure 15.6, and the new models of the upper solar layers, where the solar variable magnetic field plays a prime role, may greatly help solar-physics scientists in this job.

For a sailcraft, variable TSI affects trajectory. The closer one goes to the Sun, the higher is the TSI impact. If one swings by the Sun, the trajectory arcs approaching the Sun may extend from some weeks to a few months; since there are daily TSI fluctuations, even of 3 to 4 W/m^2, the outbound trajectory arc profile changes with respect to that with no TSI change, depending on the flyby perihelion value. This should apply in particular to the velocity direction. If the sailcraft deviates of one arc-minute at 1 AU, this translates in a miss distance of 1.75 millions of kilometers at 40 AU. If one is trying to fly by Pluto or another object of the Kuiper belt, this error could means the partial failure of the mission. One may object that deviations like this one (or even greater) could be corrected. However, a sailcraft receding from the Sun after a flyby is endowed with very high speed; it could take less than 2 weeks to pass Earth orbit, depending on perihelion and lightness number values. In other words, time for correction is short while the solar pressure decreases rapidly. This is only one of the several trajectory error sources influencing the trajectory of a sailcraft after a solar flyby. Beside the obvious attitude control uncertainties, we have to know how the space environment changes the thermo-optical properties of the sail's reflective and emissive films. Thrust acceleration and sail temperature depend on these properties as we shall see in Chapter 16.

In any case, as solar sailing is a continuous photon propulsion mode, the irradiance from the external source(s) has to be modeled accurately for any mission. This means that when a mission design is performed, one should do a sensitivity analysis, also to address our ignorance in predicting TSI in future years. The current (2007) sailcraft missions under investigation at NASA and ESA are in their preliminary stages; what now matters chiefly is the technology demonstration of solar sailing.

Another potential problem regarding TSI is its claimed isotropy. At the moment, we have no long-period measurements of TSI at high heliographic latitudes. The time series of Figure 15.6 are measurements substantially on the ecliptic, namely between -7.25 and 7.25 degrees with respect to the solar equator. From studies in recent years, most of the TSI fluctuations appear to be explained by the change of luminosity in sunspots and faculae. With respect to the mean photosphere temperature, faculae are brighter and sunspots are darker. Thus, when the number of sunspots increases, the solar irradiance is augmented. Now, among many sunspot properties, one has observed that sunspots occur in two belts symmetric with respect to the solar equator and located at $\pm(5\text{–}35)$ degrees of heliographic latitude. In addition, the sunspot umbrae (the darkest part of the sunspot) are lower in height with respect to the mean photosphere level. If a sailcraft is designed for exploring the solar poles from above, it is not taken for granted that TSI, as seen from high heliographic latitudes, varies as we measure from satellites orbiting either about Earth or about the L1 point of the Earth–Sun system.

Now let us discuss the scaling factor in Equation 15.2. We know that a general point-like source of natural light emits spherical waves. As a result, if we observe the Sun far enough, we can write $f(R, \mathbf{p}) = 1/R^2$ (where R is expressed in AU), namely, the scaling factor depends on the distance. What about if the sail is sufficiently close to the Sun? We have two combined effects.

The first consists of a reduction of the solar irradiance with respect to the $1/R^2$ law caused by the finite size of the Sun as observed from the sailcraft at a distance R. As a point of fact, each elementary area of the Sun casts its light onto the sail according to the area-sail direction. Even assuming that the Sun radiance is uniform (actually, it is not so), there is a spread of the radiation impinging on the sail. This reduces the solar pressure compared to the inverse-square law. In 1989, C.R. McInnes and J.C. Brown published $f(R, \mathbf{p})$ for a perfectly reflecting at-rest flat sail oriented radially. In 1994, author Vulpetti calculated the effect on an arbitrary-oriented moving flat sail. Although the general solution is in closed form, it is notably complicated, and its proof is quite beyond the scope of this book. The former solution is a particular case of the latter one. The main feature of such irradiance reduction is that the deviation from the ideal R^{-2} law is negligible above 0.1 AU (or about 21.5 solar radii). In particular, at 0.2 AU, such deviation is also lower than the irradiance reduction caused by the sunlight aberration on the sail.

The second reduction in solar pressure can be seen by Figure 15.7, showing two images of the photosphere in the visible band. One can immediately note that the observed solar radiance is not uniform over the solar disk. In particular,

1. the solar brightness decreases from the center to the disk edge or the limb, and
2. radiation tends to redden as the observer progressively looks toward the limb.

The overall phenomenon is referred as *limb darkening* and addresses the solar photosphere as a whole. As a point of fact, as the Sun is not uniform in its properties (in particular, temperature is height-dependent), for a number of emission spectral lines one may observe the opposite of the above points, that is, limb brightening. The contribution of such lines to the effective TSI is negligible for our purposes.

Describing limb darkening quantitatively is not a simple task, as the local and general properties of the external layers of the Sun are quite complicated. However, if one assumes thermodynamic equilibrium and solar emissivity constant over the whole electromagnetic spectrum, the so-called graybody approximation, one can get a particularly simple expression of the total radiance as function of the observer's zenithal angle as follows:

$$L(\theta) = L(0) \frac{3\cos\theta + 2}{5} \tag{15.3}$$

where $L(0)$ denotes the total radiance measured along the line of sight. There are other more accurate models of limb darkening. For instance, in recording images of the photosphere via special solar telescopes such as the solar bolometric imagers, the following model is used for limb darkening:

$$L(\theta) = L(0)\ (a_0 + a_1 \cos\theta + a_2 \cos^2\theta + \ldots) \tag{15.4}$$

where the actual number of terms is found via regression analysis.

With regard to the combined effect from finite size and limb darkening onto a general-attitude sail, some simplified formulas can be found in Chapter 4 of NASA/CR 2002-211730, June 2002, which can be downloaded from http://www.giovannivulpetti.it/.

When a deployed sailcraft orbits Earth for weeks or months, depending on its lightness number, its trajectory is further perturbed by the radiation emitted from Earth. Also, if ever a sail-based transportation system were operational back and forth between the L1-L2 points close to the Moon, then the sail motion would be perturbed by the lunar radiance. In both these cases, irradiance on the sail depends also on the changing Earth/Moon phases the sailcraft sees along its orbit. In the much more complicated case of sailcraft orbiting Earth, one has to take into account global cloud coverage and the thermal emission from much different sources such as continents and oceans; if a sailcraft spirals for long time, the different components (which can be calculated as function of time) of the irradiance should be averaged; but this task should be accomplished after the basic thrust validation in a real flight.

As a simple exercise, we can compute a rough approximation of the Earth-caused irradiance on a sail at the geostationary altitude:

1. Let us consider the zenith emission via Earth's cloud- and aerosol-free

atmosphere approximation. Earth surface, as a whole, may be assumed to be a blackbody emitter at 288 K (its radiance peak, at 10 μm, lies in the infrared region). In the near and thermal infrared regions, combining radiance and transmittance gives an interesting result: although the atmosphere transmission exhibits a complex "indented" behavior and swings many times between almost zero (full opacity) and almost 1 (full transparency), nevertheless the only range with appreciable energy emission to the outer space is 8 to 13 μm. Such a window encompasses 31.2 percent of the total surface emission, whereas we can assume roughly 60 percent of mean atmospheric transmittance. Thus, if the sail is at a distance r from the Earth barycenter ($r = 6.61$ earth radii in this example), the thermal irradiance on the sail amounts to

$$I_{\mathrm{IF}} \cong \sigma T_{earth}{}^{4}\, \xi_{8-13}\, t_{\mathrm{IF}} / r^{2} = 1.67\ \mathrm{W/m^{2}} \tag{15.5}$$

where σ, ξ_{8-13}, and t_{IF} denote the Stefan-Boltzmann constant, the power fraction emitted in the 8 to 13-μm range, and the mean atmospheric transparency in the same range, respectively. Expression 15.5 neglects the emissions from the different layers of the atmosphere and their related absorptions before exiting the full atmosphere.

2. Let us turn to the visible and near-infrared light (V-NIF or the 0.4 to 4-μm range) originating from the Sun and reflected by Earth. Although things are very complex, we can use the observed quantity known as the *planetary albedo*. Let us summarize. *Bond albedo* is defined as the total energy reflected from an object on which solar light impinges. If the object is a planet, then one gets the planetary albedo. For Earth, sunlight reflection depends strongly on water, snow/ice, vegetation, desert, clouds, human settlings, and so forth. In addition, Earth's albedo changes with latitude and regional mean surface temperatures (that affect the snow/ice extensions). When Earth's albedo is averaged over latitude/longitude, height from ground, and over one year, one gets the mean planetary albedo (here denoted by A_{p}), a useful quantity indeed. The V-NIF–related irradiance onto a sail can then be approximately by

$$I_{\mathrm{V\text{-}NIF}} \cong \tfrac{1}{4} A_{p}\, \xi_{0.4-4}\, \mathrm{TSI}/r^{2} = 3.53\ \mathrm{W/m^{2}} \tag{15.6}$$

where TSI = 1366 W/m^2 (which is the mean value of the smoothed data in Fig. 15.6), $A_{\mathrm{p}} = 0.31$, and $\xi_{0.4-4} = 0.91$ (the fractional solar irradiance at 1 AU in the V-NIF range); such values come from observations. The factor $\frac{1}{4}$ stems from the maximum cross section on spherical surface ratio.

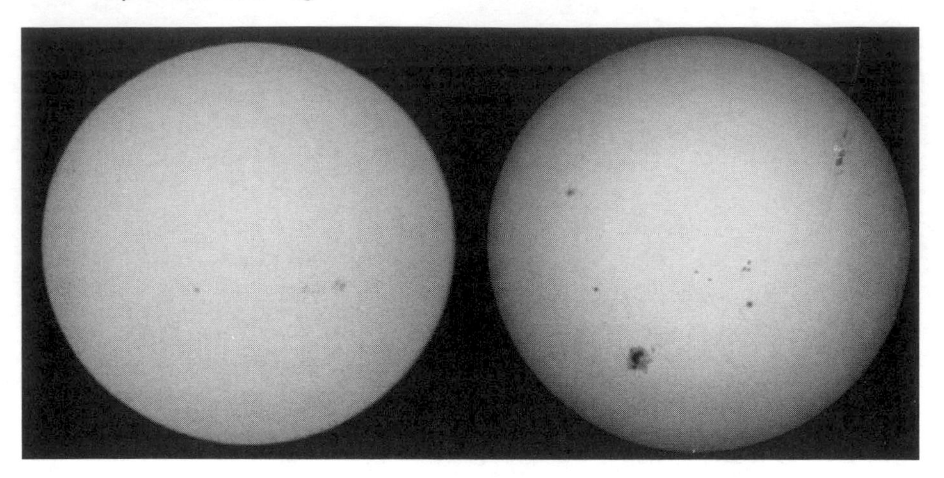

Figure 15.7. Pictures of the solar photosphere showing sunspots and the limb-darkening effect. (Courtesy of NASA) (See also color insert.)

Thus, at 1 AU from the Sun, the terrestrial irradiance onto a sail at a geostationary radius amounts to 5.2 W/m^2, according to this approximate model. This is a factor 5.7 of the mean TSI cycle amplitudes observed since 1978 (Fig. 15.6).

We have to note that the inverse-square law used in Equations 15.5 and 15.6 is not fully correct. For shorter distances, the finite-size of Earth intervenes heavily in the computation of the irradiance. The full calculation is considerably more complicated than the solar case; as a matter of fact, as the sail draws closer to Earth, it can "distinguish" the various regions (including the atmosphere) of the planet with their own radiative characteristics. The changing attitude of a sailcraft orbiting the Earth entails that, alternatively, the frontside, the backside, or both sides of the sail are irradiated in complex configurations. When a real experimental sailcraft is launched and deployed around our planet, then we may test whether the full irradiation model is sufficiently correct and how to improve it via the orbit determination process. Figure 15.8 shows the globe on 2 days: July 20 and November 20, 2006. The reader can, even at a glance, realize how complicated the radiation from Earth's surface and atmosphere may be.

Finally, let us note that the solar irradiance changes according to Earth's orbit radius. As the osculating Earth orbit has a mean eccentricity of 0.0167, irradiance at the top of Earth's atmosphere varies from 1321 and 1413 W/m^2. Such change affects Equations 15.5 and 15.6 the same way; as a point of fact, $\sigma T_{\text{earth}}{}^4$ is proportional to TSI.

We have shown how delicate the matter of computing the irradiance on the sail area is. We note again that this is the amount of radiation received by

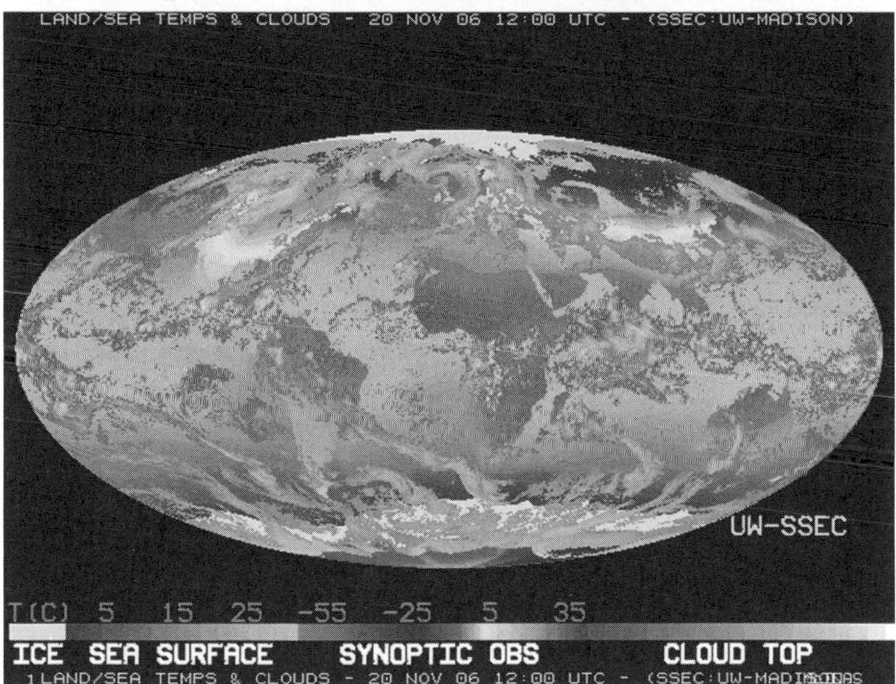

Figure 15.8. Globe geostationary satellite composite images for 2 days in 2006. (Courtesy University of Wisconsin at Madison) (See also color insert.)

the sail *before* it interacts with photons. Chapter 16 describes such an interaction.

Acknowledgments

We thank Dr. Robert F. Cahalan, head of the Climate & Radiation Branch of NASA's Goddard Space Flight Center, for his *Mathematica* notebook *SolarIrr-3.nb* on solar and terrestrial irradiances, on which we based Figures 15.2, 15.4, and 15.5. Special credit goes to Physikalisch-Meteorologisches Observatorium, Davos, Switzerland, designated to serve as a world radiation center since 1971, for the composite TSI time series, shown in Figure 15.6.

Further Reading

http://climate.gsfc.nasa.gov/.
http://www.pmodwrc.ch/.
http://umbra.nascom.nasa.gov/sdac.html; http://sohowww.nascom.nasa.gov/.
http://www.gigahertz-optik.com/database_en/html/applications-tutorials/.
http://www.optics.arizona.edu/Palmer/rpfaq/rpfaq.htm.
http://www.solarmonitor.org/index.php.
http://www.albedoarts.net/Define.html.

Modeling Thrust from Electromagnetic Radiation Pressure 16

In Chapter 15, we saw what type of light the Sun and Earth cast into space, each body with its own characteristics. In particular, we emphasized that 30 years of high-precision measurements from satellites have revealed that the total solar irradiance fluctuates, and which parts of its spectrum can affect a solar sail significantly. In practice, the wavelength range from 50 nanometers to 20 microns contains all solar energy flux in which we are interested.

Nomenclature

In this chapter, capital and lower case **bold** letters denote vectors in the usual three-dimensional space.

Frames of Reference

It is time to address the problem of the interaction of the solar photons with sail materials. To keep things simple but meaningful, we use a model where the sail's macroscopic behavior is highlighted. In other words, we model the vector thrust acceleration induced by the solar photons impinging onto the sail. What matters is the acceleration sensed onboard, and its components with respect to some sailcraft-centered reference frame. Once this vector acceleration is established, one can transform it into the reference frame where the spacecraft motion is described. (Actually, this is a general principle for computing the trajectories of any spacecraft.)

We need two reference frames here: the heliocentric inertial frame (HIF), and the sailcraft orbital frame (SOF). The HIF has its origin in the Sun's barycenter (SB). To define HIF (and other reference frames), we need a reference epoch, which is January 1, 2000, 12:00:00 Terrestrial Time (TT). This date can be expressed equivalently as January 1, 2000, 11:59:27.816

G. Vulpetti et al., *Solar Sails*, DOI: 10.1007/978-0-387-68500-7_16,

International Atomic Time (TAI), or January 1, 2000, 11:58:55.816 Coordinated Universal Time (UTC), namely, the everyday time scale. Normally, such date is denoted by either J2000.0 or J2000. TT may be thought of as an ideal form of TAI. TT and TAI differ only by an offset: TT = TAI + 32.184 seconds. This bias is inessential for our aims here.

Let us choose an HIF having the mean ecliptic and equinox at J2000 as its reference plane and its x-axis, respectively. The y-axis and z-axis follow consequently. With respect to HIF, let **R** and **V** denote the sailcraft's instantaneous vector position and velocity, respectively.

Note 1: Actually a frame centered on SB cannot be inertial in the strict sense, since the Sun slowly revolves about the barycenter of the solar system (SSB). In principle, the SB–SSB distance ranges from 0 to about 2 solar radii (1 solar-radius = 696,000 km), with a mean value slightly more than 1 solar radius. This shift induces extra accelerations to the interplanetary spacecraft, which are smaller for shorter distances from the Sun. Unless one wants to compute trajectories with very high accuracy, normally one can assume that HIF is truly inertial.

Note 2: The time scales that one uses either in everyday life or for scientific purposes are very important and should not be undervalued. The attentive reader is invited to begin with an introduction to this complex area of scientific investigation by visiting the web site http://en.wikipedia.org/wiki/Time (and the many references therein). Then, for in-depth scientific understanding, we recommend http://www.iers.org/, http://physics.nist.gov/Genint/Time/time.html, and http://tycho.usno.navy.mil/.

Let us define SOF. This is a spacecraft-centered frame. We have to distinguish two cases:

1. If the sailcraft flight begins with a *direct* (or counterclockwise) motion, as observed in HIF, then the x-axis is determined by the unit vector $\mathbf{r} = \mathbf{R}/R$, whereas the reference plane is given by the plane determined by **R** and **V** (assumed nonparallel to one another). The z-axis is given by the direction **h** of the sailcraft's orbital angular momentum per unit mass, or $\mathbf{H} = \mathbf{R} \times \mathbf{V}$.
2. If the sailcraft flight begins with a *retrograde* (or clockwise) motion, as observed in HIF, then the x-axis is the same as in case 1, but the z-axis is oriented *opposite* to **H** and, consequently, the y-axis is in the semi-plane $(\mathbf{R}, -\mathbf{V})$.

This definition of SOF, which extends the classical concept of orbit frame, reflects the fact that a general sailcraft trajectory may be composed of direct and retrograde arcs separated by at least one point where the orbital angular momentum either vanishes in magnitude ($H = 0$) or in its Z-component

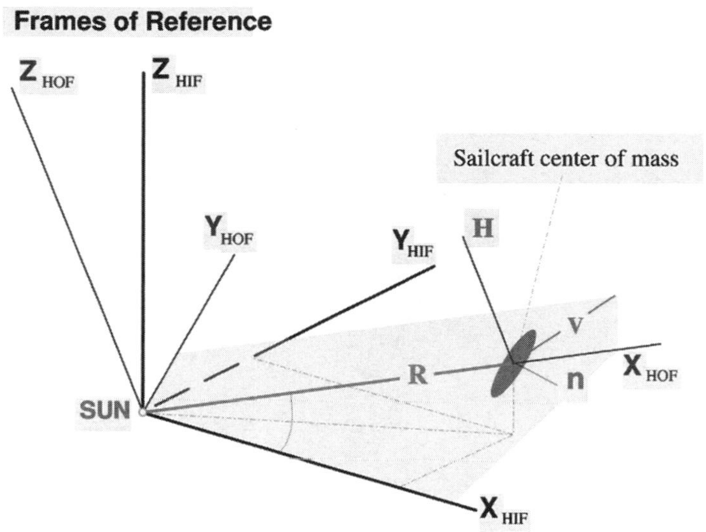

Figure 16.1. HIF and SOF frames of reference (see text).

$(H_Z = 0)$ only. It is possible to show that SOF always evolves *smoothly* with respect to HIF. Therefore, there exists a continuous time-variable orthogonal matrix, which transforms SOF vectors into HIF vectors, and vice versa. Figure 16.1 sketches both reference frames (SOF is counterclockwise) together with the position, velocity, and angular momentum of a sailcraft. (We translated the SOF axes to the Sun and renamed them HOF for a better graphic view.)

Because we are dealing with classical (i.e., nonrelativistic) dynamics, HIF and SOF are sufficient to characterize the interplanetary sailcraft motion. In this framework, HIF and SOF can share the same time scale, which we assume to be the above-mentioned Terrestrial Time.

Phenomena Transferring Momentum

Now we have the complete (i.e., including time scale) reference frames and can model the vector thrust. Figure 16.2 shows a circular flat sail arbitrarily oriented in SOF. We are choosing the orientation unit vector **n** (orthogonal to the sail) contained in the semispace opposite to the reflective layer. In such semispace, **n** can be resolved in SOF via two angles, the azimuth α and the elevation δ, defined quite similarly to longitude and latitude. Therefore:

$$\mathbf{n}^{\mathrm{T}} \equiv (n_{\mathrm{x}}\ n_{\mathrm{y}}\ n_{\mathrm{x}}) = (\cos\alpha\ \cos\delta \quad \sin\alpha\ \cos\delta \quad \sin\delta) \qquad (16.1)$$

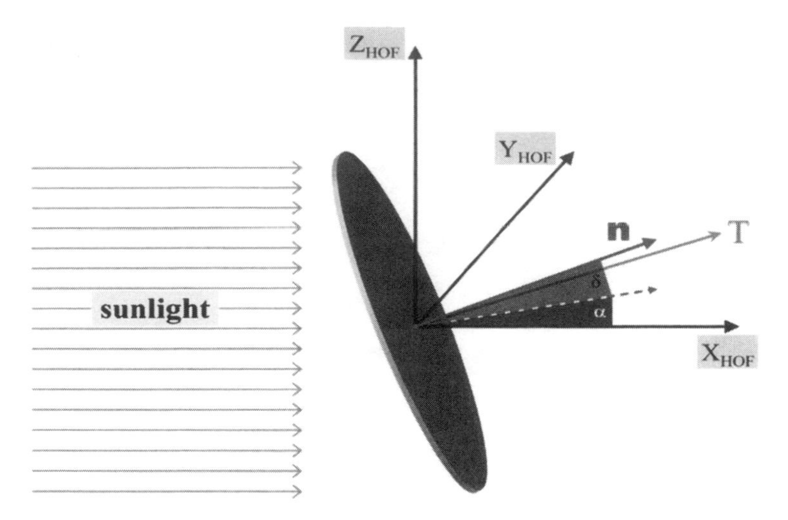

Figure 16.2. Scheme of the thrust of a sail in SOF. Note that thrust is not along the sail's normal line, but always in the plane angle delimited by the sail axis **n** and the sunlight direction **u** (or the direction of $\mathbf{X}_{SOF}$). When sail axis is aligned with sunlight, all three directions coincide.

We can proceed with describing the effects of the interaction between photons and sail. In general, the solar-sail materials exhibit specular reflection, diffuse reflection, and absorption of sunlight. We will not choose material types and thicknesses that allow sunlight to be transmitted. Light may be partially transmitted through metals if their thickness is sufficiently small, but no thrust would be produced by such photons if the plastic substrate were transparent. Reflection, diffusion, and absorption of photons are complex quantum-physics phenomena indeed. Here, we will use a very simplified picture, stressing those macroscopic properties of interest in solar sailing.

Specular reflected light causes a net force normal to the sail surface and oriented as **n**. This is very easy to check in SOF by using the momentum conservation law applied to photons and sail (which is at rest here) just before and after the interaction.

Then, let us consider *absorption*. Because absorbed photons are not re-emitted, they transfer their momentum to the sail body in the same direction of incidence. In SOF, such direction is simply the unit vector $\mathbf{u}^T \equiv (1 \;\; 0 \;\; 0)$, namely, the instantaneous radial direction of sunlight. However, the energy absorbed by the sail's reflective layer is subsequently emitted by both sail sides. This emission not only determines the sail's mean temperature, but also induces a further net thrust. For understanding how, we have to note that each side emits in its semispace (2π sr): the light emitted in any solid elementary angle causes recoil momentum. Such momenta have to be summed to get the total momentum acting on the body. If the considered surface side were Lambertian, then the radiance would be constant (by definition) and the total recoil momentum rate would be directed along the direction normal to the opposite side. Its magnitude would be exactly two thirds of the momentum rate that would result if the side has the same radiant exitance (Chapter 15) parallel to its orientation. A real surface has an emission/diffusion coefficient, say, χ different from two thirds, in general. Combining the net effect from both sail sides, the net thrust results to be proportional to the absorbed power and to the following dimensionless factor:

$$\kappa \equiv \frac{\chi_f \, \varepsilon_f(T) - \chi_b \, \varepsilon_b(T)}{\varepsilon_f(T) + \varepsilon_b(T)} \tag{16.2}$$

where ε denotes the temperature-dependent emittance of the surface and T is the sail temperature. The subscripts refer to the frontside and backside of the sail. Note that whereas χ_f and χ_b may be almost equal to one another, the emissive layer of the sail exhibits an emittance much higher than the reflective layer's, just for lessening T drastically. Consequently, κ is negative in general; namely, this thrust contribution is directed along $-\mathbf{n}$.

Now, let us consider the light diffused by the reflective layer. When a beam of light impinges onto a real surface at some given angle, the reflected light consists of two parts: one is the direct or specular reflection, and the other one is the diffuse reflection. A surface never is perfectly smooth and may be pictured as a random sequence of very small hills and valleys with respect to the mean surface level. On the reflective layer, such irregularities spread the re-emitted light over the hemisphere; the recoil momentum distribution is characterized by the coefficient χ_f, which generally indicates a nonuniform radiance. For simplicity, one assumes that the net momentum rate has azimuthal symmetry, so that this thrust component is along $\mathbf{n}$. Surface roughness produces a reduction of thrust magnitude with respect to what is expected from a full specular reflection.

However, the story is not complete. From a microscopic viewpoint, diffuse reflection takes place in two steps. First, photons are absorbed by the surface molecules. Second, according to the characteristic times of the molecules, photons may be re-emitted along a random direction, even if all incident photons had the same direction; this step has been characterized above. The first step transfers momentum to the sail parallel to the incidence direction, which is represented by $\mathbf{u}$ in SOF, just like the absorption.

All in all, we can assert that a sunlight beam impinging on a reflective sail produces a thrust consisting of two main components: the more intense is

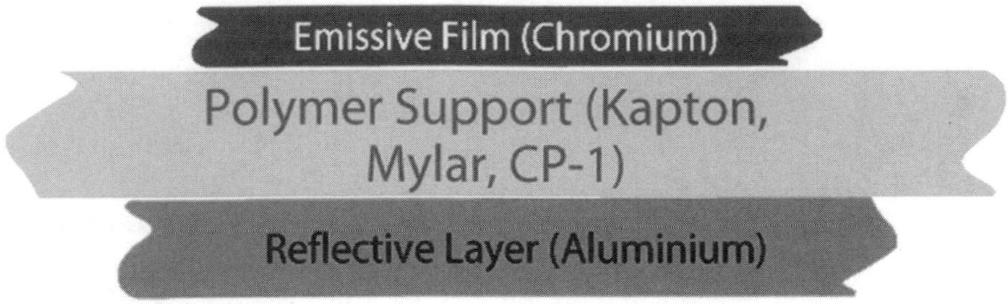

Emissive Film (Chromium)

Polymer Support (Kapton, Mylar, CP-1)

Reflective Layer (Aluminium)

Figure 16.3. A typical sail's multilayer arrangement. Aluminium and chromium are deposited on a plastic substrate, which may consist of Kapton, CP-1, or mylar for the first solar-sail missions currently under investigation at NASA and ESA.

directed along **n**, and the other one is directed along **u**. (Equivalently, one could split the total thrust parallel to **n** and along the sail surface. However, **n** and **u** both have a direct physical meaning.) There are other phenomena affecting the actual thrust of sunlight on a metal layer, but those already described are the most important for a *flat* sail.

We will report the thrust acceleration equation here without proof (what was said previously is sufficient to verify the main aspects of the equation). We suppose that the sail materials configuration is like that sketched in Figure 16.3, namely, a three-layer film. If the sail is sufficiently far from the Sun ($R > 0.1$ AU, in practice), then in SOF:

$$\mathbf{A}^{\mathrm{SOF}} = g_{\odot}\,\mathbf{L} = \left(\frac{GM_{\odot}}{R^2}\right)\left(\frac{1}{2}\frac{\sigma_c}{\sigma}\right)\{n_x[(2r_{\mathrm{spec}}n_x + \chi_{\mathrm{f}}\,r_{\mathrm{diff}} + \kappa a)\mathbf{n} + (a + r_{\mathrm{diff}})\mathbf{u}]\} \equiv \mathrm{B}\Gamma \tag{16.3}$$

where we have set

$$B \equiv \frac{GM_{\odot}}{R^2}\,\frac{1}{2}\,\frac{\sigma_c}{\sigma}\,n_x$$

$$\sigma_c \equiv 2\,\frac{I_{1\mathrm{AU}}}{c g_{1\mathrm{AU}}} \cong 1.5368\ \mathrm{g/m^2}\ (I_{1\mathrm{AU}} = 1366\ \mathrm{W/m^2}) \tag{16.4}$$

$$\sigma \equiv m/S$$

where $\mathbf{L} \equiv (l_x\ l_y\ l_z)^{\mathrm{T}}$ denotes the lightness vector, namely, the thrust acceleration normalized to the local solar gravitational acceleration ($g_{\odot}$) and resolved in the orbital reference frame. On the right side of Equation 16.3, the first parentheses contain the gravitational acceleration magnitude to be multiplied by the semi-ratio between the critical loading (σ_c) and the actual sailcraft (sail) loading (σ), namely, the ratio between the whole sailcraft mass m and the effective sail area S.

Normally, one writes $GM_{\odot}/R^2 = g_{1AU}(\frac{AU}{R})^2 = 0.005930\ [m/s^2]/R[\text{AU}]^2$ with the Sun-sailcraft distance R expressed in AU. (Note that the mean solar gravity, which keeps Earth in its stable orbit in spite of the perturbations from the other planets, amounts to less than 6 mm/s^2, that is 0.6 thousandth of the mean Earth gravity on ground.)

The quantity in braces is a dimensionless vector whose magnitude can vary in the interval $[0, 2]$. The set $\{r_{\text{spec}}, r_{\text{diff}}, a\}$ represents appropriate mean values of the specular reflectance, diffuse reflectance, and absorptance, respectively, of the sail frontside. Finally, "c" denotes the speed of light we mentioned in the previous parts of this book.

Note 3: The relationship $l\sigma - \tau\sigma_c = 0$ is a useful identity between $|\mathbf{L}| \equiv l$ (or the lightness number), the sailcraft loading (σ), and the concept of thrust efficiency (τ), that is, the ratio of the actual thrust on the thrust of a perfect at-rest sail orthogonal to sunlight.

The components of **L** can be named the *radial* (lightness) number, the *transversal* number, and the *normal* number, respectively. The full mathematics related to **L** is beyond the scope of this book; however, we will shortly mention that (1) a sailcraft can gain/lose energy only if $l_y \neq 0$, (2) its orbital angular momentum can be changed in direction only if $l_z \neq 0$, and (3) its orbital angular momentum can be changed in magnitude only via $l_y \neq 0$ again. Such properties are of fundamental importance for sailcraft mission design.

Thrust Acceleration Features

Now we are able to highlight the many features the sailcraft acceleration equation exhibits. First, let us see what the term B represents. Assume that we are building a sail with the same material and properties on both sides. Additionally, this sail is supposed to behave as a blackbody, in practice. The first assumption entails $\kappa = 0$; the second one implies $a = 1$. As a result, it is very easy to carry out

$$\mathbf{A}^{\text{SOF}}|_{\text{blackbody}} = \left(\frac{GM_{\odot}}{R^2}\,\frac{1}{2}\,\frac{\sigma_c}{\sigma}\,n_x\right)\mathbf{u} = B\mathbf{u} = \frac{GM_{\odot}}{R^2}\,(l_x\ 0\ 0)^{\text{T}} \qquad (16.5)$$

In words, B represents the magnitude of the acceleration experienced by a *blackbody* sail. The magnitude of this acceleration depends on the sail attitude through n_x. However, its direction always is radial outward, no matter how the sail may be oriented. That limits considerably the set of trajectories such a sailcraft is able to run. On the other hand, a transparent sail would produce zero acceleration since plainly $r_{spec} = r_{diff} = a = 0$. Thus, we can state the basic property that a solar sail with full *controllable* thrust requires materials having $r_{spec} + r_{diff} > 0$. In this context, two limit cases may be distinct. If the reflectance were only diffuse, then one could get

$$\mathbf{A}^{SOF}\big|_{diffusion} = B(\chi_f\, \mathbf{n}+\mathbf{u}) \tag{16.6}$$

If the sail were perfectly specular, then the acceleration would equal

$$\mathbf{A}^{SOF}\big|_{full\text{-}reflection} = 2B\, \mathbf{n} \tag{16.7}$$

Equations 16.3 to 16.7 tell us that the direction of the thrust acceleration always lies in the plane angle $\widehat{\mathbf{nu}}$, bounds included, as shown in Figure 16.2.

With regard to the magnitude of $\mathbf{A}^{SOF}$, this can be increased by decreasing σ and increasing the specular part of the total reflectance. (Theoretical studies and experimental work are in progress for building a sail with all the properties required by a real space mission.) In general, $\mathbf{A}^{SOF}$ changes during a flight simply because the sail attitude has to be varied for driving the sailcraft to the target. The maximum value of $|\mathbf{A}^{SOF}|$ at 1 AU is defined as the characteristic acceleration. Although this sole value does not determine the sailcraft trajectory class, it is often used for comparing two missions of the same class; it is a useful parameter, but one should remember that only the time-dependent vector function $\mathbf{L}(t)$ defines completely the trajectory between the starting point and the desired target. Trajectory classes can be defined through **L**'s components exhaustively.

The features so far discussed focus on the main behavior of thrust. Things are less simple in reality. By using the above concepts, let us try to grasp what should happen in practice. The lightness vector components (defined above) can be obtained by varying the attitude angles α and δ. One may be induced to think that the various lightness numbers can be found by solely inserting Equation 16.1 into Equation 16.3. Actually, the thermo-optical parameters of the sail's frontside and backside are not constant throughout the solar spectrum. In particular, reflectance mainly depends on wavelength and polarization of light, surface roughness, and the incidence angle of sunlight onto the reflective layer. In the case of a reflective layer of aluminium with a thickness greater than ~ 100 nm, Figures 16.4 and 16.5 show the *averaged* behaviors of specular reflectance, diffuse reflectance, and absorptance as a function of the incidence angle and the surface roughness. Note that high roughness values cause r_{spec} reduction around the normal incidence. The opposite holds for r_{diff}. In processing experimental data, we have utilized the theory of scalar scattering for separating the specular and scattered contributions to the total reflectance. With regard to the absorptance, it may be considered constant up to incidence angles of about

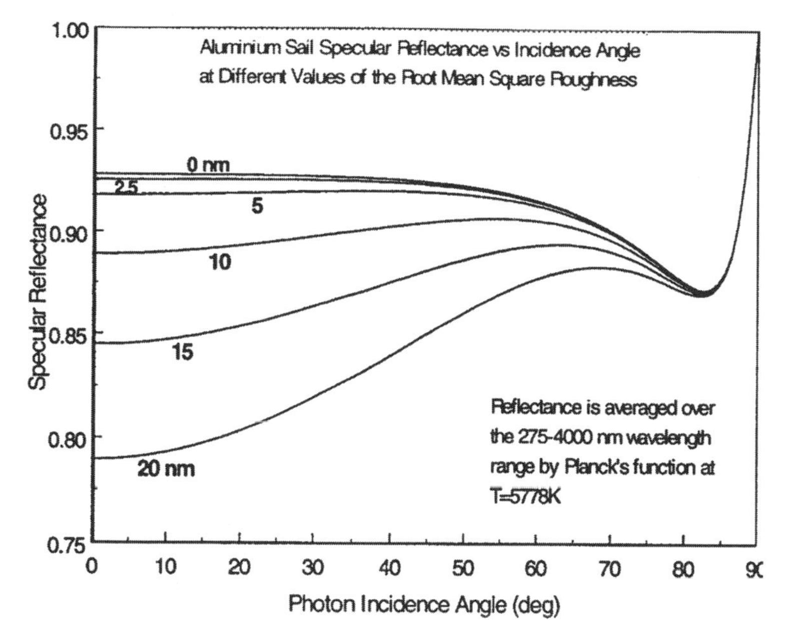

Figure 16.4. Thin aluminum-film *specular* reflectance plotted as function of the sunlight incidence angle and surface roughness. (From G. Vulpetti and S. Scagline, Further Reading)

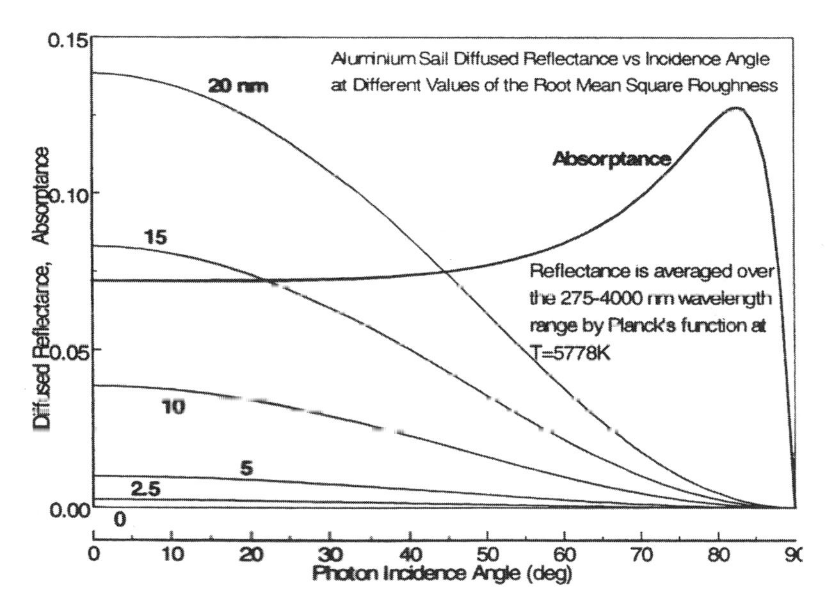

Figure 16.5. Thin aluminum-film *diffuse* reflectance shown as a function of the sunlight incidence angle and surface roughness. Absorptance is a function of the incidence angle. (From G. Vulpetti and S. Scaglione, Further Reading)

35 degrees. Then, it increases and achieves a local maximum at 83 degrees. Note that absorptance does not depend on roughness.

Note 4: in designing any sailcraft mission, theoretical models about the thermo-optical parameters of the sail sides and extended measurements should be used for making an accurate trajectory profile. However, all such calculations have to be repeated as soon as the mission analyst receives the profiles of the thermo-optical properties *measured on the sail to be actually flown*. This point is quite important—conceptually and practically—especially for the first real sailcraft mission. Subsequently, once the orbit determination process is activated as part of the operations at the mission control center, the necessary refinements to the thrust model can be accomplished.

Behavior of the Thrust Acceleration Components

Figure 16.6 shows the lightness vector x-y-z components that we named the radial, transversal, and normal lightness numbers, respectively, for a sailcraft with σ=10 g/m^2 and at 1 AU from the Sun. The TSI has been assumed to be 1366 W/m^2, very close to the average over the last three solar cycles. The lightness numbers are plotted versus sail azimuth between 0 and 90 degrees for different nonnegative values of sail elevation, both in SOF. Figure 16.7 displays **L**'s behavior as function of the full azimuth range for two symmetric values of elevation. The same scales have been kept in all subplots for ease of comparison. A number of interesting behaviors can be inferred from both figures, as follows:

1. The relative shapes of the **L**'s components are independent of the particular σ-value, as expressed by Equation 16.3.
2. The radial number always is nonnegative and exhibits a bell-like shape with the maximum at $\alpha = 0$.
3. The transversal number shows one local maximum and one local minimum, which are independent of the sail azimuth, in practice. In addition, one gets $sign(l_y) = sign(\alpha)$. (When σ goes below 2.1 g/m^2, this property can be utilized for fast solar sailing.)
4. The normal number is either nonnegative or nonpositive depending on the sail attitude elevation, namely $sign(l_z) = sign(\delta)$. In both cases, it looks like a wide bell with either maximum or minimum at $\alpha = 0$.
5. For a large range in elevation, the normal number is rather lower than the radial number; however, if elevation is sufficiently high, then it can overcome the radial number.

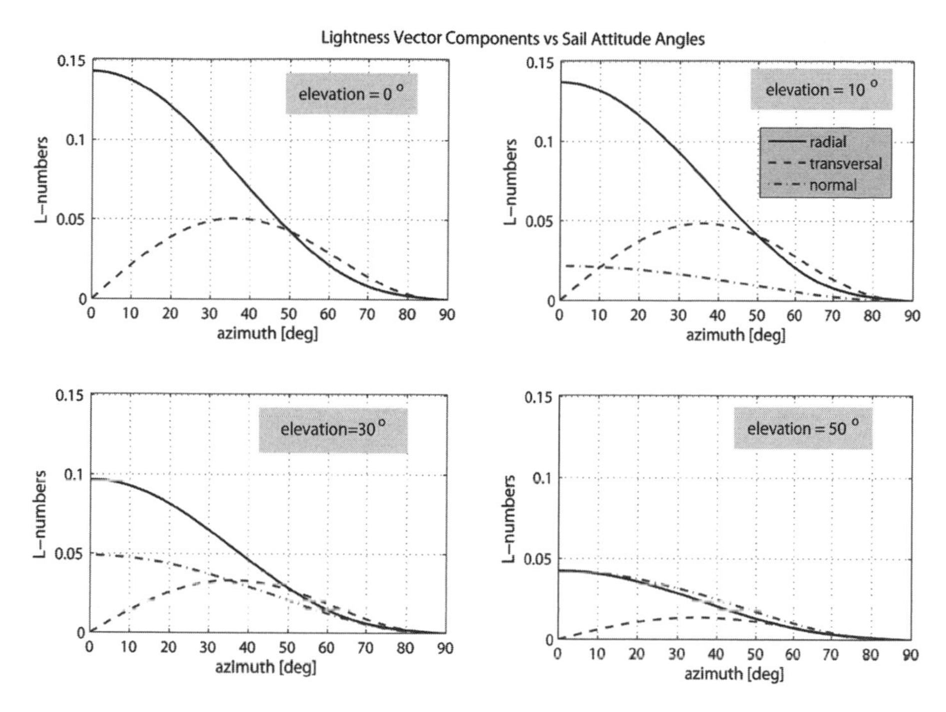

Figure 16.6. Components of the lightness vector for different sail attitudes. Lightness numbers are shown as function of the sail azimuth for various sail elevations. TSI has been assumed equal to 1366 W/m^2 and the sailcraft located at 1 AU from the Sun. Sailcraft sail loading was supposed to be equal to 10 g/m^2. Sail reflective-layer root mean square roughness amounts to 20 nm.

6. The transversal number crosses the radial one and becomes higher, but its value is of the same order of the radial one.
7. At very high/low azimuth values, the radial number is the highest one, but all components all go to vanish at $\alpha = \pm 90$ degrees.

Property 7 holds for flat sails or sails with sufficiently low temperatures. However, since the sail's substrate is of plastic material, the sail progressively billows as the sailcraft draws closer to the Sun. In such case, acceleration can vanish when the sunlight incidence angle is slightly more than 60 degrees.

Finally, we end this chapter by mentioning a few issues about the influence of the thermo-optical parameters degradation on the sailcraft acceleration and structures. Equations 16.6 and 16.7 show that roughness is a sort of *intrinsic* degradation in terms of thrust acceleration because of the different coefficients that the specular and diffuse terms have in Equation 16.3. However, there is an overall optical *external* degradation caused by

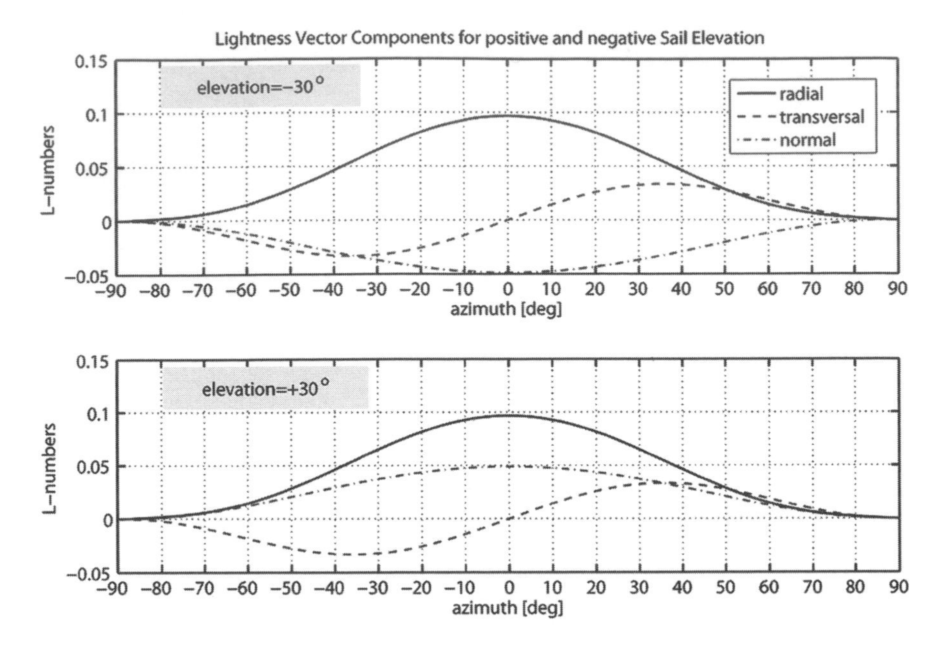

Figure 16.7. Components of the lightness vector versus the full admissible range in azimuth for positive and negative elevations. The other parameters are the same as in Figure 16.6.

the space environment. In fact, (mainly) solar-wind protons and ultraviolet photons, both emitted by the Sun abundantly, modify the surface of the sail's reflective layer. Such changes in the layer lattice increase with time and depend on the energy of the particles impinging onto the sail surface. Thus, the longer a sailcraft remains in space close to the Sun, the higher the change of the thermo-optical parameters. Such a change is permanent, meaning that if a sail were tilted at zero thrust at some time during the flight, then the original sail state could not be recovered. On a conceptual basis, one can note that surface roughness causes part of the (otherwise specular) reflectance to turn into diffuse reflectance; in contrast, when external degradation is active, part of the total reflectance turns into absorptance. This can increase the sail temperature and decrease the available thrust significantly, depending on the mission. For instance, a mission surveying the solar poles at a fraction of AU should be carefully designed even from the viewpoint of the operational orbit, which progressively gets closer and closer to the Sun.

Thrust Acceleration in the Heliocentric Inertial Frame

The orthogonal matrix that transforms a vector, resolved in SOF, into the corresponding vector with components in HIF is given by

$$\Xi = (\mathbf{r} \;\; \mathbf{h} \times \mathbf{r} \;\; \mathbf{h}) \tag{16.8}$$

This agrees with the SOF definition given in §2. Therefore, it is immediate to get

$$\mathbf{A}^{\mathrm{HIF}} = \Xi\, \mathbf{A}^{\mathrm{SOF}} = g_{\odot}\, (\mathbf{r}\; \mathbf{h}\times\mathbf{r}\; \mathbf{h}) \begin{pmatrix} l_x \\ l_y \\ l_z \end{pmatrix} = \frac{g_{1\mathrm{AU}}}{R^2}\, (l_x\, \mathbf{r} + l_y\, \mathbf{h}\times\mathbf{r} + l_z\, \mathbf{h}) \tag{16.9}$$

(where R is expressed in AU). This vector has to be summed to the solar gravitational acceleration and any other perturbation acceleration vectors; altogether they are equal to the time derivative of the sailcraft velocity.

Further Reading

http://en.wikipedia.org/wiki/Solar_sail.

http://en.wikipedia.org/wiki/Fast_solar_sailing.

http://nuke.giovannivulpetti.it/SolarSailing/tabid/56/Default.aspx (documents and links).

Colin R. McInnes, *Solar Sailing: Technology, Dynamics and Mission Applications*, Springer-Praxis, Chichester, UK, **1999**.

G. Vulpetti and S. Scaglione, T*he Aurora Project: Estimation of the Optical Sail Parameters*, Second IAA International Symposium on Realistic Near-Term Advanced Scientific Space Missions, June 29–July 1, **1998**, Aosta, Italy. Also, special issue of Acta Astronautica, vol. 44, no. 2–4, pp. 123–132, **1999**.

Sailcraft Trajectories 17

We are nearing the end of this introductory book on solar sailing. We saved one of the most intriguing topics—trajectory design—for last. However, it is beyond the scope of this book to delve deeply into mathematics and the related physical aspects. So after a very short presentation of the sailcraft motion equations, we discuss the class of trajectories (and missions) via several technical plots. Some trajectories have been designed in past decades, some were investigated in the first years of this century, and some have been calculated specifically for this book by means of modern (and very complex) computer codes. This chapter discusses sailcraft motion equations in their simple form, using no additional mathematics; presents generalized Keplerian orbits that only sailcraft can draw; describes interplanetary transfer by solar sailing; describes some of the new striking features solar-sail propulsion offers, such as the possibility of designing orbits that differ from the Keplerian ones significantly, which allows a mission designer to move beyond the limits of conventional spacecraft; discusses the behavior of a sailcraft under the gravitational influence of more than one celestial body; highlights the so-called artificial equilibrium points; explains the high nonlinear feature of very low sail-loading sailcraft.

Motion Equations

Formally, the classical motion equations of spacecraft are not complicated. But in addition to the difficulty, common to any scientific discipline, of modeling the real world, in astrodynamics there is the need to optimize trajectories with respect to some performance criterion related to the mission goals. Furthermore, one has to solve equations numerically, which may seem a trivial task, at first glance, in the modern era of high-level software and computers. However, the opposite is true. Judicious choices of calculation units, reference frames, switching from one frame to another (when applicable), numeric integrators, optimization methods, and

G. Vulpetti et al., *Solar Sails*, DOI: 10.1007/978-0-387-68500-7_17,

computing additional information (useful for understanding the many aspects of the problems) are all delicate procedures for achieving reliable results upon which a mission may be designed. There is an incredible amount of high-level literature on these topics. We cite a few studies in the course of this chapter relatively to the sailcraft trajectories we are discussing.

In the inertial reference frame (IF) where one wants to describe the sailcraft motion, let **r** denote the position vector of the sailcraft, namely, the vector from the origin of IF to the sailcraft's barycenter. Often, the IF origin coincides with the center of mass of a celestial body in the solar system (the Sun or any planet), which is taken as the central body. Let $GM \equiv \mu$ indicate the gravitational mass of such a *central* body. Then, the equation of motion can be written as follows:

$$\frac{d^2\mathbf{r}}{dt^2} = -\mu_{\ast}\mathbf{r}/r^3 + \boldsymbol{P} + (\mu_{\odot}/R^2)\,\Phi\,\mathbf{L} \tag{17.1}$$

where the subscript ✷ stands for the Sun (⊙), Earth (⊗), or any other planet. Let us explain the meaning of this vector equation. Like any classical motion equation, the left-hand side (l.h.s.) represents the vector acceleration (a pure kinematical quantity). The right-hand side (r.h.s.) includes all dynamical contributions, namely, the forces per unit mass that act upon the space vehicle. The first term on the r.h.s. represents the gravitational vector acceleration due to the central body; $\boldsymbol{P}$ denotes the overall perturbation acceleration stemming from conservative or nonconservative fields (but propulsion), whereas the third term is the solar-sail thrust acceleration due to the sunlight's solar pressure, as described in other chapters of this book. R is the distance from the sailcraft to the Sun, **L** is the sailcraft lightness vector defined in Equation 16.3 in Chapter 16. Finally, Φ is the rotation matrix from HOF (the reference frame defined in Chapter 16) to IF. Such a matrix transforms **L**'s components in HOF to those in the current inertial reference frame. The degradation of the sail's reflectance (Chapter 18) cannot be included in an equation of this type. However, as discussed in Chapter 16, one can include varying-with-sunlight incidence reflection and absorption in this equation.

Therefore, for our purposes, the above equation of motion can be considered general. Of course, we have to specify either initial conditions (the initial value problem) or mixed information regarding the initial and final state (the two-boundary problem) of the sailcraft in order to integrate such second-order differential equation *numerically*. However, that is not enough. Like any differential equation containing free parameters, or *controls*, Equation 17.1 needs a time profile of such controls in order to be solved completely. How may we specify such control behaviors? Normally, in designing a trajectory (or a set of trajectories) for a space mission with given payload goals, one has to identify *three* important classes of linear/nonlinear constraints and some objective function: (1) state equality/inequality constraints, (2) control equality/inequality constraints, and (3) mixed state and control constraints. Classes 1 and 2 in particular are characteristic of the set of admissible trajectories, whereas class 3 is the direct consequence of the optimization problem at hand. Designing space missions based on rocket propulsion has a strong feature: minimizing the propellant during the transfer-to-target phase and during the operational mission (orbit and attitude control). In contrast, sailcraft-based space missions will be characterized—among the many features we have highlighted in the previous chapters—by two noteworthy aspects: (1) minimizing the transfer flight time, and (2) achieving controlled orbital configurations. These aspects often are impossible to accomplish via rocket propulsion. Optimization criterion (1) is important not only for cutting mission costs and a number of equally significant nontechnical issues, but also for reducing the degradation of the reflective layer of the sail, another significant objective indeed.

Equation 17.1 may contain the contribution to non-Keplerian acceleration coming from the pressure of the light backscattered or emitted by a planet or a natural satellite (where applicable). Because *planetary*-radiation thrust acceleration is always much weaker than *solar*-radiation thrust acceleration (due to a planet temperature that is very low compared to the solar one, the planetary albedo, considerable absorbing and scattering planet atmosphere, etc.), nobody thinks of using the planet's radiation for controlling the trajectory of a sailcraft. Instead, this acceleration—even if modeled in computer codes (at least in the most sophisticated ones)—may be dealt with as a perturbation to sailcraft spiraling about a planet. Therefore, it can be considered as included in the term $\boldsymbol{P}$ of Equation 17.1, which mainly includes gravitational perturbations, caused by celestial bodies other than the central body and its aspherical shape.

General Keplerian Orbits

The best way to see the effects of solar-radiation thrust is to analyze a number of very special trajectory classes by first removing the perturbation term from Equation 17.1 and considering heliocentric sail trajectories. In addition, let us suppose that the sail direction is always parallel to the local sunlight direction. From the general equation, one gets

$$\frac{d^2\mathbf{R}}{dt^2} = -(1-l_x)\frac{\mu_\odot}{R^3}\mathbf{R} \tag{17.2}$$

where l_x is the radial lightness number defined in Chapter 16. This equation is the differential equation governing all possible *general* Keplerian orbits. Conceptually, they are very simple: the sailcraft senses the Sun with an *effective* gravitational mass $\tilde{\mu}_\odot = (1-l_x)\,\mu_\odot$. Depending on the technology that we will utilize progressively in future solar sailing, $\tilde{\mu}_\odot$ may become negative too; that is, the sum of the solar gravity and the solar-radiation effect can be *repulsive* on the sailcraft. In the special case that the radial number is identically one (that is, $\sigma = \tau\sigma_c$ from Note 3 in Chapter 16), the sailcraft can move uniformly on a rectilinear trajectory, with speed and direction depending on the initial conditions, in the solar system. (Perturbations alter this ideal state, of course.) This property lends itself to intriguing interplanetary transfers. Such an advanced-sail and spacecraft technology would allow sailcraft to spiral fast (1 month at most) about Earth and to escape the Earth–Moon system with a residual speed of 1 to 2 km/s. Thus, the heliocentric speed may amount to 6.5–6.7 AU/year. The velocity direction is approximately perpendicular to the position vector of the Earth, and in Earth's heliocentric orbit plane, at the exiting time. (More generally, such property holds for any sailcraft with $l_x = 1$, and leaving any planet.) After a rectilinear arc, the sail would be tilted so as to progressively match the heliocentric vector velocity of the target planet (e.g., Mars). If we include the spiral time about Mars, the whole Earth-to-Mars transfer should last half the time of the current flights to the red planet. However, the main advantage would be that most of the heliocentric trajectory would be orthogonal to the departure-planet vector position, as stated above, with the immediate consequence of relaxing the launch window considerably; in other words, going to Mars would not need to wait for favorable Earth–Mars relative positions. A similar thing would apply for the return trip. Thus, a reusable sailcraft may accomplish round-trips to Mars twice as fast and with low dependence on the planetary positions! Since the sail technology for such flights is somewhat advanced with respect to that available today, missions like this one have not yet been studied carefully. (At the time of this writing, preliminary research is in progress in Italy that is investigating the possibility of using special carbon nanotube membranes for future applications to solar sailing, though we are not yet able to estimate how far in the future all this may take place).

Other interesting novel solutions happen if $0<l_x<1$. They are the subject of the rest of this section. Let consider a heliocentric circular orbit obeying

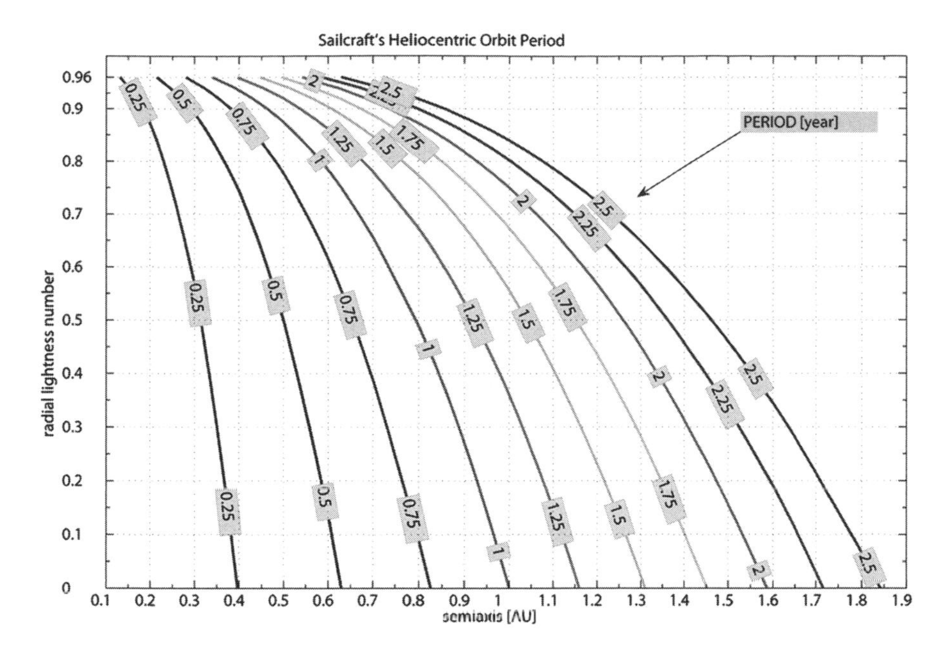

Figure 17.1. Contours of the period [year] of a general Keplerian orbit as a function of the semimajor axis [AU] and the radial lightness number.

Equation 17.2. What is the period of such an orbit? Figure 17.1 shows period values as a function of the orbit radius and the radial number.

Note 1: The circular-orbit solutions to Equation 17.2 correspond to orbits with the sail *already* deployed. If the spacecraft initially orbits about the Sun circularly with the sail unfolded and subsequently opens its sail radially, then the orbit transforms into an elliptic, parabolic or hyperbolic, depending on the radial number. The sailcraft pre-deployment speed has to be lower than the usual circular value $\sqrt{\mu_{\odot}/R}$ for getting a sail-open circular orbit. For instance, the deployment maneuver can happen at the aphelion of a pure Keplerian ellipse.

In Figure 17.1, note the curve representing the 1-year period orbits. In particular, one can envisage a sailcraft on the same plane of Earth orbit and 0.3 AU sunward and always on the same Sun–Earth line. (This is analogous to what happens close to the Sun–Earth system L1 point; however, L1 is 1.5 million kilometers or 0.01 AU far from Earth). This sail mission concept requires $l_x = 0.657$, namely, a technology considerably more advanced than the current one. Such mission could be not only scientific but also utilitarian. Among other things, as anticipated in Chapter 9, the space-storm warning time would range from 16 to 31 hours instead of the current 70 to 90 minutes from L1 (including the lower-speed path of the solar wind in Earth's magnetopause).

Implicit in Note 1 is that Figure 17.1 holds for any elliptic orbit of a sailcraft about the Sun. We invite college students to start from Equation 17.2 and carry out some general formulas (e.g., energy, angular momentum, eccentricity, semimajor axis, etc.) useful to evaluate the performance of a sailcraft with respect to a classical spacecraft capable of rocket maneuver. As a point of fact, the sail deployment may be viewed as an impulsive maneuver with no propellant consumption. For a spacecraft orbiting about the Sun on any circular path, deploying a sail—completely sunward and with lightness number equal to ½—inserts the vehicle into a parabolic orbit. The open interval (1/2, 1) entails a hyperbolic orbit with some excess velocity.

With the concepts established so far, we are able to apply a sail mission concept studied years ago by author Matloff. The following is an ideal mission from some viewpoints; nevertheless, it will be useful for defining the concepts of fast and very-fast solar sailing modes, which we will deal with later (see Fast and Very Fast Sailing). Suppose that a spacecraft has two propulsion systems: (1) an impulsive rocket engine, and (2) a photon solar sail. Some launcher delivers the vehicle to about 2.7 million kilometers from Earth where the solar field dominates. Let us assume that the rocket is capable of releasing a velocity impulse equal to $(\sqrt{2}-1)/\sqrt{R_0}$ (or 12.3 km/s at $R_0 = 1$ AU, very large indeed). Were such impulse applied parallel to Earth's orbital velocity at that point ($\mathbf{V}_E$), the spacecraft would escape the solar system on a parabola. One knows that in such a case the speed at infinity is $V_\infty = 0$. However, let us apply the same impulse antiparallel to $\mathbf{V}_E$. The ensuing orbit is elliptic sunward and the vehicle can achieve the perihelion $R_P = \frac{1}{2}(\sqrt{2}-1)\,R_0$ (or 0.207 AU) after about 85.5 days (during which the rocket system has been jettisoned). Here, the sail comes into play. Let us suppose (1) we deploy the sail in a very brief time interval and radially, having designed the sailcraft with $l_x = 1$, (2) to keep the sail radial from that moment on. The scenario, which we may call the *CRS* (combined rocket-sail) reference flight, is depicted in Figure 17.2. The ensuing motion is rectilinear and uniform with a speed equal to twice the parabolic speed at R_0, namely, 84.2 km/s or about 17.8 AU/year. This value is almost five times the speed of Voyager 1, the fastest spacecraft launched hitherto.

The analysis performed above is as simple as it is striking: specific values have been reported, but the result about the cruise speed depends only on the initial circular orbit. Two observations:

1. To achieve a final high speed, the spacecraft first has to lose most of its initial kinetic energy, about 65.7 percent, independently of the initial

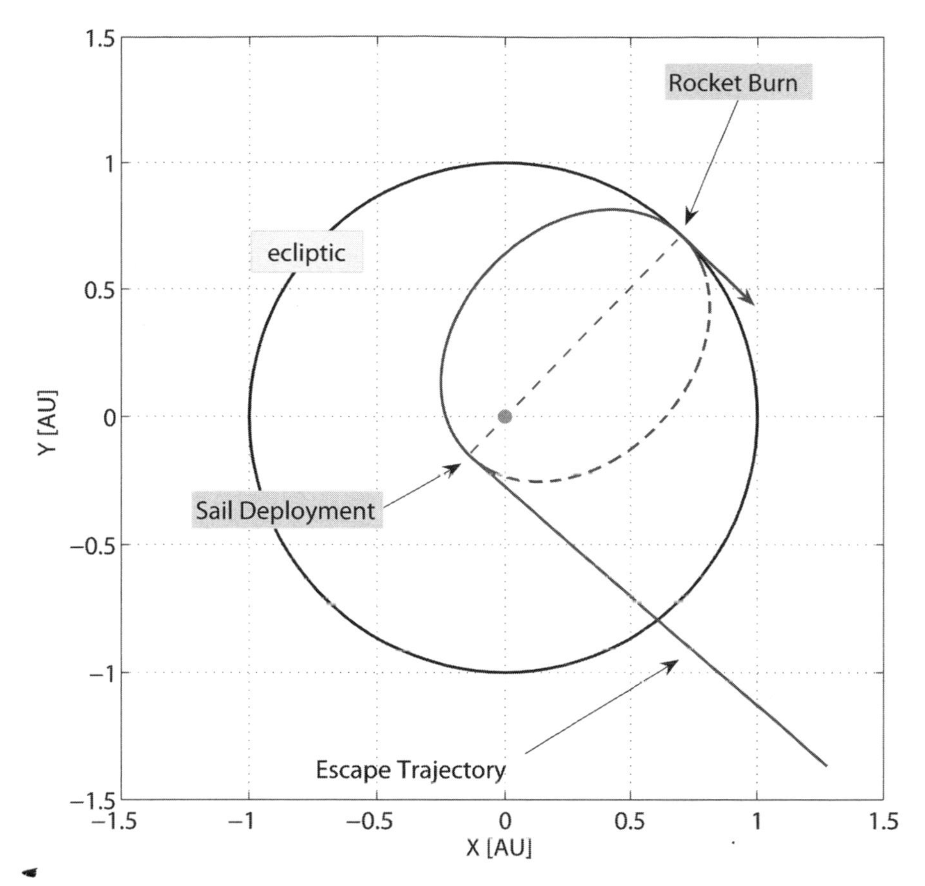

Figure 17.2. Ideal *two-propulsion* reference mission. An impulsive rocket burn is applied opposite to the orbital velocity at R_0. At the perihelion of the elliptic orbit, the sail is deployed impulsively, radially, and giving the sailcraft a lightness number equal to 1. This generates an escape rectilinear trajectory with a cruise speed equal to twice the parabolic speed at R_0.

orbit. Subsequently and closer to the Sun, the radiation pressure will be able to give the sailcraft a much higher energy.

2. If one wants to achieve 84 km/s of excess speed (cruise speed) by a single rocket (tangential) impulse at 1 AU (where V_{circ} =29.785 km/s), from the energy equation, it is straightforward to carry out ΔV=64.2 km/s, a huge impulse, a "mission impossible" for rockets.

Much probably, the above mission concept will be unfeasible for a number of reasons that are beyond the scopes of this book; nevertheless, it is meaningful from a theoretical viewpoint. As a point of fact, we will consider again such scenario later (see Fast and Very Fast Sailing), when we discuss realistic high-speed sailcraft trajectories.

Interplanetary Transfers

General Keplerian orbits are based on sailing always orthogonal to the local sunlight direction as sensed in HOF. The general control of a sail is expressed via the lightness vector in the vector Equation 17.1. We discussed the **L**-vector and the sailcraft acceleration components in Chapter 16. Here, and qualitatively, we summarize the dynamical role of the lightness components in order to introduce the reader to the interplanetary transfers. Then, we will show a number of meaningful examples of them.

Apart from gravitational and solar-wind perturbations from solar system bodies, a heliocentric sailcraft undergoes three fields when the sail is arbitrarily oriented in HOF: (1) the Sun's local gravity, (2) the *radial* sunlight-pressure force component, and (3) the *orthogonal* sunlight-pressure force component. Fields 1 and 2 are conservative; in contrast, field 3 is nonconservative. It is a strange behavior due to field splitting. From Chapter 16, it is possible to prove the following sailcraft features:

1. The orbital energy, though depending on l_x, can be increased/decreased only if $l_y \neq 0$.
2. The angular momentum can be changed in direction only if $l_z \neq 0$.
3. The angular momentum can be varied in magnitude only via $l_y \neq 0$ again.

The utilization of these properties allows the mission analyst to design any interplanetary transfer flight. Of course, a mathematical algorithm of optimization, with equality and inequality constraints, produces a set of locally optimal trajectories. The problem of which is the best one for a certain mission depends on the characteristics of the whole project one is dealing with. (This is the most difficult astrodynamical problem to be solved for each planned mission.)

Most interplanetary transfers are of the rendezvous type. Although such rendezvous transfers are from the heliocentric orbit viewpoint only (i.e., the gravitational fields of the departure and arrival planets have not been considered), they are of significant historical meaning. The departure and the arrival may be approximately viewed as position-velocity states on the so-called spheres of gravitational influence. Furthermore, to illustrate the main trajectories properties, some flights have been supposed to be coplanar with the planetary orbits (this is not true, of course, but this approximation

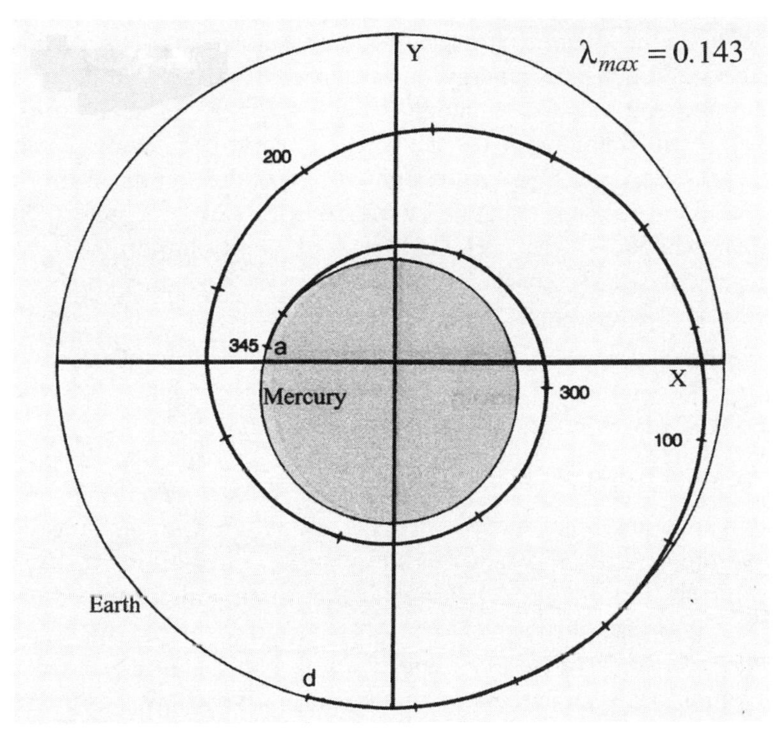

Figure 17.3. Example of Earth-to-Mercury transfer via solar sailing. Departure date was June 7, 1987. The maximum lightness number (but not achieved in the trajectory profile) is 0.143.

allows one to focus on the principal properties of the transfer trajectory). Nevertheless, the authors have re-computed such profiles in the modern view expressed in Chapter 16. One of the most meaningful novelties is the introduction of variable thermo-optical parameters, such as those shown in Figures 16.4 and 16.5 in Chapter 16. When appropriate, we will comment on such results.

Figure 17.3 shows the 1987–88 transfer from Earth to Mercury. As is apparent, the sail has to be kept at a negative azimuth in HOF for slowly decreasing its distance R from the Sun. For over 3 months, the sailcraft speed V decreases slowly as well; then, although the sail attitude angle continues to be negative, V begins increasing because of the lower R. In the last month of the transfer, the sail azimuth is positive for progressively matching the orbital velocity of Mercury. The total transfer lasts 345 days. An interesting three-dimensional transfer from the real Earth orbit to the real Mercury orbit can be computed for the 2020–2021 opportunity. A combination of transfer time and perihelion of the transfer trajectory can be optimized to 382 days/0.33 AU, respectively, with only three simple attitude maneuvers in HOF.

Apart from the attitude strategy, one should note that the sailcraft has to be transferred in orbital energy from -0.5 to -1.292, while its orbital angular momentum has to change from 1 to 0.609, in solar units. Such changes are enormous with respect to the current propulsion capabilities. Let us digress for a moment. An enlightening example comes from the next 2-spacecraft cornerstone mission, by the European Space Agency (ESA) and the Japanese Space Agency (JAXA), named BepiColombo, for exploring Mercury deeply. BepiColombo should be set off in 2013 on a journey lasting about 6 years! The transfer from Earth to Mercury is rather complicated. After launch into a geostationary transfer orbit, the Mercury composite spacecraft will first use chemical propulsion. Then, the spacecraft will be set on its interplanetary trajectory through a Moon flyby. On its way to Mercury, the spacecraft will brake against the Sun's gravity. BepiColombo will accomplish this by utilizing of the gravity of Earth, Venus, and Mercury itself and by using solar electric propulsion. When approaching Mercury, BepiColombo will use the planet's gravity and a conventional rocket engine to insert itself into the target polar orbit. Like many other important NASA and ESA missions, the astrodynamical realization of the flight depends completely on the relative positions of some planets; in other words, the launch window is very narrow, its repeatability is very low, and the transfer time is long. The authors have checked that, if sailcraft technology with a lightness number of 0.16 were ready in 2013 (it will not be), then the real heliocentric transfer to Mercury would last 367 days. Mission opportunities would repeat more than one in every year and in different months, since there is no flyby need for gaining delta-V. Earth escape could be performed by a launcher providing a small hyperbolic excess speed, whereas Mercury capture (via solar sailing) may be aided by a small-size chemical engine. To within a quarter of a century (or less), Earth–Mercury–Earth regular round-trips may become a reality for many scientific missions.

Figure 17.4 shows two Earth-to-Venus rendezvous transfers, in 1981–83, lasting about 204 and 213 days, respectively. Sailcraft technology has been assumed to perform $\lambda_{max} = 0.168$, or 1 mm/s^2 as characteristic acceleration. In each profile, most of the trajectory is a decelerating arc, and then the final orbit matching occurs via a slight accelerating arc. For the future, we have computed a good 218-day opportunity with departure on December 6, 2019 (using the same technology). As opposed to missions to Mercury, good launches repeat every 20 months; transfer times can be different essentially because of the Venus orbit inclination (about 3.4 degrees) over the ecliptic, and Earth and Venus orbit nodes (which differ by about 100 degrees, on the average).

Figure 17.5 shows an example of Earth-to-Mars heliocentric rendezvous

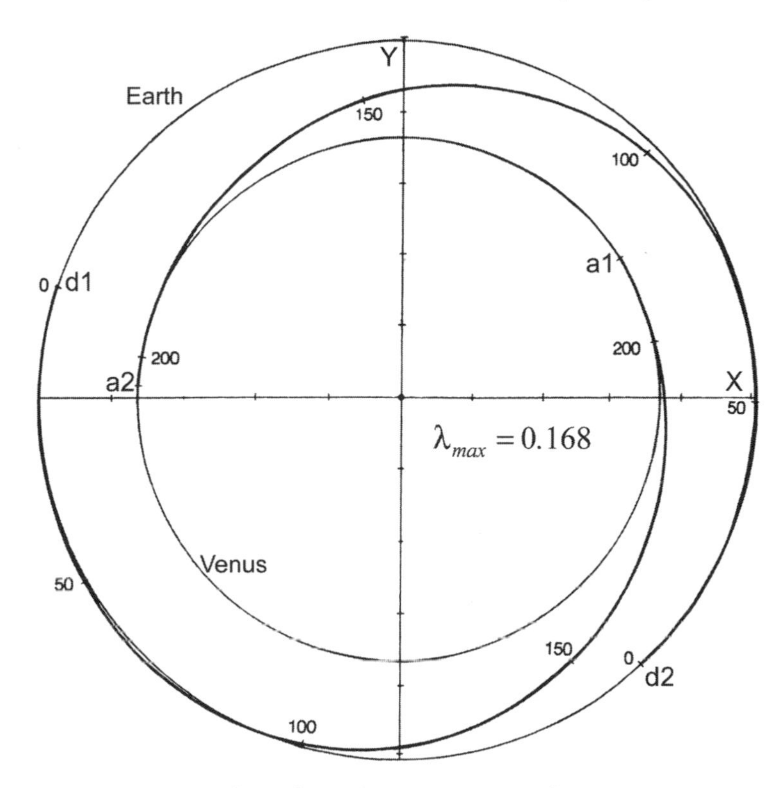

Figure 17.4. Two examples of Earth-to-Venus transfer via solar sailing with maximum lightness number equal to 0.168. Departures were on August 5, 1981 and March 4, 1983.

transfers in 2033, as computed by the authors. (Earth and Mars orbits are those corresponding to 2033–2034). Let us at once say that there are many, many opportunities of rendezvous with Mars in the next two decades. We have chosen an overall "medium-technology" sailcraft with a total sail loading of 10 g/m^2 and a 425-day journey to Mars. This σ-value is about one order of magnitude better than the ESA Geosail mission (in 2011–2013). The small squares in Figure 17.5 mark the departure, the arrival, and two intermediate attitude maneuvers. Thus, the minimum-time attitude control is piecewise-constant in HOF, and is simple to implement. One can suppose that even using a moderate sailcraft technology as a whole, a rather large sail could be designed realistically in the 2030s. Thus, if a 1-km^2 sail could be managed, a 10-tonne (1 tonne = 1 metric ton) ferry sailcraft would go forth and back between Earth and Mars, most probably with a net payload of 7 tonnes; in that period, many infrastructures on the Mars surface would be built. If the ferry sailcraft works, NASA, ESA, and JAXA may have their

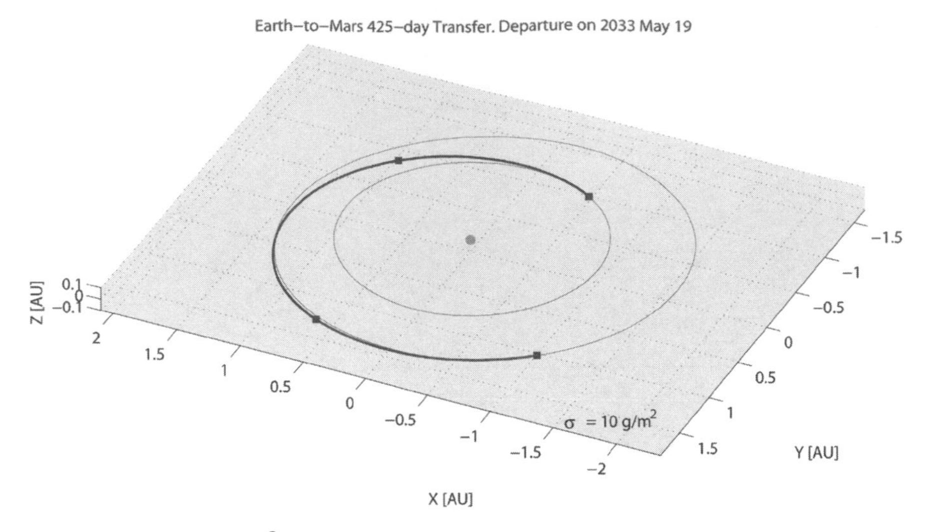

Figure 17.5. A 10-g/m^2 sailcraft minimum-time trajectory to Mars. Departure is on May 19, 2033; arrival is on July 9, 2034. Squares denote departure, arrival, and intermediate attitude maneuvres. The three trajectory arcs correspond to three HOF-constant attitudes.

ferries for Mars colonization. Of course, the set of trajectories of ferry sailcraft is more complex than what we have shown here; some of the related concepts hold, though.

Sailcraft can go to the planets beyond Mars. However, we do not present trajectory profiles here because the celestial-mechanics concepts of inner and outer spheres of influence could not be neglected for the distant large planets of the solar system. Only a few examples can be found in the specialized literature, but they refer to sailcraft "entering" some unstated sphere of planetary influence with too high a speed for ignoring a number of problems. The design of even simple sailcraft trajectories to the outer planets is beyond the scope of this chapter.

We showed three figures of significant historical importance; the trajectory that a large sailcraft—envisaged, studied, and fostered by the Jet Propulsion Laboratory (JPL)—would have followed to rendezvous with the Halley comet in 1986. However, NASA headquarters did not approve the mission, and the history of solar sailing changed. However, who knows? Perhaps, if that very complex mission failed for some unexpected reason, public opinion would have been that the mission was unimpressive, and now neither NASA nor ESA would be engaged in solar sailing, the authors of this book included!

Figure 17.6 shows the three phases of the full trajectory to Halley. JPL

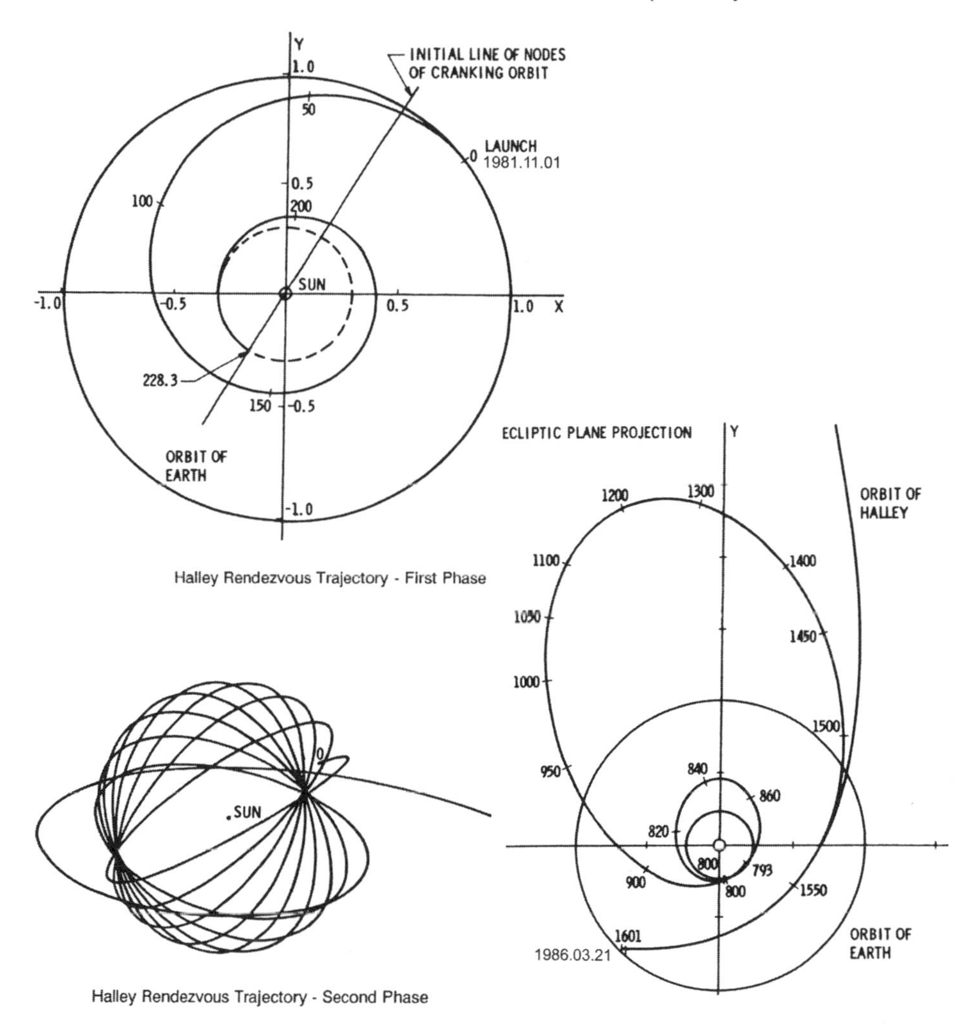

Figure 17.6. Halley comet rendezvous trajectory as designed by the Jet Propulsion Laboratory in 1970s. The conceived sail was a square of 800 m on each side; sailcraft sail loading was 7.7 g/m^2, and the maximum lightness number was equal to 0.177. (A) Two phases of the full transfer to Halley. (B) Retrograde trajectory arc to arrive at the comet post-perihelion rendezvous. (Courtesy of NASA)

conceived a square sail of 800 m on each side, allowing a sailcraft sail loading of 7.7 g/m^2 and maximum lightness number equal to 0.177. After a ballistic injection into the solar field (at 3.5 km/s) and sail deployment, the sailcraft is controlled such to spiral down and move to a circular orbit at 0.25 AU, but inclined 20 degrees with respect to the mean plane of Earth's orbit (see *ecliptic* in the Glossary). This completes the first transfer phase, as plotted in Figure 17.6 (top left). In the second phase, shown in Figure 17.6 (bottom left), the sailcraft maneuvers in attitude and changes the inclination of its osculating orbits until about 145 degrees are achieved. As we know, orbits with inclination higher than 90 degrees are retrograde. As a point of fact, not only is Halley comet strongly inclined with respect to the ecliptic, but also its motion is opposite to that of the planets, namely, it is clockwise. The sailcraft cranking in this phase can be described as a circular orbit slowly rotating about an axis lying on the ecliptic (i.e., the axis indicated at the top of the figure). The third phase consists of a slow increase of energy and angular momentum parallel to that of the comet, as shown in Figure 17.6 (right), due to the relatively low lightness number. The sailcraft approaches the comet from below, resulting in a post-perihelion rendezvous (a bit less than 1 AU). Then, the sail system is jettisoned so the spacecraft moves with the comet and eventually lands on it.

A few comments: First, a similar mission accomplished by some electric engines would be hugely expensive. Second, the first two phases of the above trajectory design suggest a way to send a sailcraft over the solar poles, namely, in an orbit literally orthogonal to the solar equator. Current concepts of sailing to the solar poles are based on such a trajectory strategy. Third, a rendezvous near the comet's perihelion (~0.58 AU) would be desirable for scientific purposes. Fourth, according to JPL, the next Halley's perihelion will occur on May 31, 2061. There is plenty of time for designing a sailcraft using new technology. Together with the progress already made in solar-sailing astrodynamics, all that will certainly result in a completely new, much faster, and adjustable rendezvous trajectory, and robots for exploring and probing the comet. These topics may be developed in Ph.D. theses for graduate students in astrodynamics or aerospace engineering.

Non-Keplerian Orbits

When one talks about Keplerian orbits, either conventional or generalized (see General Keplerian Orbits, above), one implicitly assumes a basic fact: the instantaneous plane of the orbit passes though the barycenter of the central body. This holds even in many-body dynamics via the concept of osculating orbit. Now, let us suppose that all components of the lightness vector are nonzero except l_y (the transversal number). According to Equation 16.9 in Chapter 16, the sailcraft thrust acceleration in HIF has no azimuthal component. In addition, according to points 1 and 3 in the list in the previous section, the sailcraft's angular-momentum magnitude and orbital energy are constant. In contrast, the angular-momentum changes in direction; this excludes any set of circular orbits, the planes of which contain the solar center of mass. Do such factors have a more direct geometrical meaning? Yes, they have. So far, the general solution to Equation 17.1 with $l_y = 0$ has not been investigated (at least to our knowledge); only a particular, but rich, class of orbits has been studied. However, by theoretical and numerical tools, it is possible to show that there exists a special class of orbits that can be simply described as follows:

$$\mathbf{R}(t) = \mathbf{s} + \mathbf{C}(t), \quad \mathbf{s} \cdot \mathbf{C}(t) = 0 \tag{17.3}$$

where **s** is a constant vector and $\mathbf{C}(t)$ is the vector position of either a circular or elliptic orbit perpendicular to **s**. In other words, we can get circles or ellipses on planes not passing through the Sun! Vector **s** is just the shift of the orbital plane. In the past years, shifted circular orbits have been well studied from many interesting viewpoints, including potential applications. In contrast, the larger class of the shifted ellipses, their existence region, stability, and control have not been dealt with appropriately (maybe because of its higher mathematical complexity). We have checked numerically that ellipses orthogonal to arbitrary shift vectors surely exist in the case of *ideal* sails. Of course, their geometrical properties are tightly related to the lightness vector and, as a consequence, to the sail technology and the evolution of the sail materials in the space environment (see Chapter 18).

The above feature holds also for planetocentric sailcraft, in particular Earth-bound. Thus, applications of shifted circular and elliptic orbits may be various. At the end of the 1980s, Robert L. Forward suggested utilizing some displaced geostationary orbit (GEO) for allowing telecommunications over Earth's poles, which cannot be seen by conventional GEO satellites just for the reason that their planes have to pass through Earth's center of mass. He made preliminary calculations and invited colleagues to pursue further studies. In 2005, author Vulpetti, for his lectures at the Aerospace School of Rome University, computed aluminized-sail operational orbits and showed the relationships between orbit displacement, orbit period, pole elevation angle, and sailcraft sail loading. A simple example of shifted GEO orbit is shown in Figure 17.7, which is slightly out of scale. This operational orbit is outside the zone of danger due to space debris (in the figure, the white dots represent a sort of visual mean distribution of space debris according to NASA). Table 17.1 contains the main parameters related to such an orbit. The reported numbers are self-explanatory. The ultra-low value of the sail loading entails a factor of 800 in technological improvement with respect to

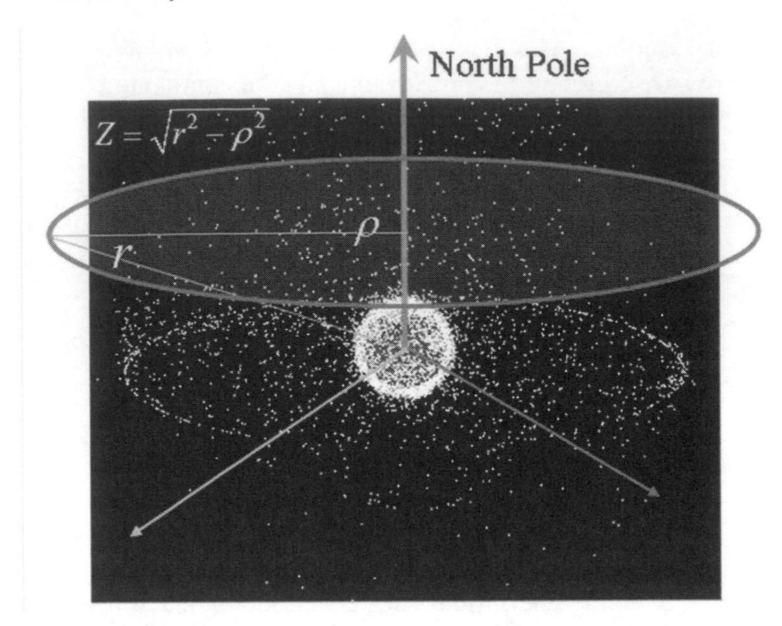

Figure 17.7. Displaced geostationary orbit for telecommunications over Earth's North Pole. (Courtesy of NASA; adapted by G. Vulpetti)

Table 17.1 Main parameters relatively to the nominal shifted geostationary orbit, shown in Figure 17.7, for observing terrestrial zones around the North Pole

Operational-orbit period	86164.0989 s (1 sidereal day)
Sailcraft elevation at North Pole	11 degrees
Orbit displacement (Z)	2.211 Earth radii
Orbit radius (ρ)	6.2300 Earth radii
Sail distance (r)	6.6107 Earth radii
Lightness number (λ) {Sun, Earth}	{12.65, 0.334}
Sailcraft sail loading (σ)	0.1215 g/m^2

the ESA Geosail mission under phase A study. It should not be possible to increase appreciably the distance of the sailcraft from Earth so that the local gravitational acceleration is significantly lower. As a point of fact, even polar-zone telecommunications imply (1) reasonable elevation angles (the sailcraft is seen from any observer on ground), and (2) the requirement that voice-telecom should not have a time delay higher than the current 0.25 seconds. The users of GEO telecom satellites have all experienced annoyance from this delay during telephone conversations especially on the job.

The analysis of realistic-sail-shifted ellipses (or quasi-ellipses) should be another topic of frontier astrodynamics, to be addressed in M.S. theses or Ph.D. dissertations.

Many-Body Orbits

The many-body problem in celestial mechanics is known not to have a general analytical solution except for the two-body and the restricted three-body problems. Nevertheless, the many-body system is well managed numerically, as shown in several successful space missions. The JPL ephemeris file DE-405 is an excellent example of that. However, even the restricted two-body solar sailing (Sun plus sailcraft), as drawn from Equation 17.1 with $\boldsymbol{P} = 0$, exhibits no general closed-form solution, even in the two-dimensional case. There exist special solutions, though, in the restricted Sun–planet–sailcraft problem, namely, two primaries plus an infinitesimal mass body sensitive to the pressure of the light emitted by either primary. These solutions can be found if one assumes that (1) the primaries (of masses M_1 and M_2) revolve on circular (coplanar) orbits about their common barycenter; (2) the sail has *specular* reflectance (which may be lower than 1, but with no diffuse component); and (3) the sail attitude is fixed in the baricentric reference frame (BRF), the frame co-rotating with the primaries. In problems of this kind, what matters is essentially finding the sailcraft's position and velocity histories in BRF.

The mathematical analysis of such a problem proceeds similarly to the well-known restricted three-body problem, with primaries moving in circles. The classical problem exhibits five equilibrium points (of which three are unstable) in the circles plane. However, in the current case, one has three more parameters: the sail loading σ or, equivalently, the lightness number (since the sail is perfectly reflecting), and the two angles specifying the sail orientation $\mathbf{n}$. In principle, σ and $\mathbf{n}$ can be chosen arbitrarily. Given a set $\{\sigma, \mathbf{n}\}$, the sailcraft has a certain *radial* number and, as discussed above (see General Keplerian Orbits), sees the luminous primary with a reduced mass $M_1{}^* = (1 - l_x)\, M_1$ (here, l_x is assumed to be less than 1). As a result, there are equilibrium points for the equivalent restricted problem with primaries $\{M^*{}_1, M_2\}$; if there is a nonzero *normal* number, then such points are expected to be displaced with respect to the common plane of the primaries' motion. Varying the parameters $\{\sigma, \mathbf{n}\}$ in a continuous range entails an infinite set of equilibrium points, which are quite *local* and *relative* in nature, namely, they are sensed only by the sailcraft. (One might call them *artificial* equilibrium points, but one should remember that different sailcraft sense different equilibria, in general). In this context, there are *allowed* three-dimensional space regions of equilibrium points *induced* by the full set of values $\{M_1 , M_2 , \sigma, \mathbf{n}\}$. If $\sigma \to \infty$ (i.e., a spacecraft without sail), then one finds the classical Lagrange points again.

What about a sailcraft that moves in the Earth–Moon system? By using the mathematical tool known as the theory of perturbations, it is possible to investigate many features dependent also on the particular mass ratio $M_{\text{Moon}}/M_{\text{Earth}} = 0.0123$ of such system. For example, *above* the classical Lagrange points, sailcraft may move on an elliptic displaced orbit, but active control is required; in other words, sail attitude has to be trimmed to compensate for secular effects.

What about a sailcraft that is placed in the Sun-Earth system ($M_{\text{Sun}}/M_{\text{Earth}} = 333{,}000$)? Perturbative analysis of sailcraft dynamics shows that the equilibrium points are generally unstable. However, due to the very small perturbation accelerations around them, a sail orientation trimming strategy should be sufficient for getting a long stay of the sailcraft around these points.

Some of the above considerations even hold for a rocket-powered spacecraft. For instance, NASA spacecraft ACE (Advanced Composition Explorer) has been orbiting around the classical L_1 point of the Sun–Earth system. This libration point orbit is unstable and four to six station-keeping maneuvers per year are required to keep the spacecraft bound to L_1. The overall fuel consumption per year amounts to about 4.1 kg (9 lb). There is sufficient fuel onboard to allow operations until 2022. However, ACE (or similar spacecraft) would be unable to find long-term equilibrium in arbitrary regions of the Sun–Earth gravitational system. Although they are conceptually possible, the fuel to be spent via any control strategy would be enormous or expensive at least. This is not the case for sailcraft, as we already know.

Colin McInnes (now at University of Strathclyde, Glasgow, UK) has studied for years halo orbits and their potential applications. They have many problems that are not yet investigated due to the huge complexity of the many-body problem. For instance, suitable research on special realistic-sail–induced equilibrium regions under the influence of the Sun and a number ($N > 2$) of planets would be high desirable. Again, this could be the subject of Ph.D. dissertations. Potential applications, not known so far, may arise.

Fast and Very Fast Sailing

Let us resume the CRS reference mission described earlier. We can distinguish two semi-open intervals of speed: (1) $[V^{\text{par}}, 2V^{\text{par}})_{|R_0}$, (2) $[2V^{\text{par}}, V^{\text{star}})_{|R_0}$. If a sailcraft endowed with a certain sail loading can be guided such a way that its cruise speed belongs to interval 1, then the sailcraft is said to perform a *fast* solar sailing. But if the cruise speed falls in open interval 2, then we talk about *very fast* solar sailing. Thus, the CRS reference flight represents a "conceptual attainment" of the (otherwise arbitrary) lower bound of the very fast range. Below, we discuss the upper bound V^{star}.

The theory of fast and very fast solar sailing is a rather complex topic of solar sailing. It was formulated only in the mid-1990s, and it has not yet been completely investigated. This intriguing research is beyond the scope of this book. Nevertheless, we will explain the basic principles qualitatively and, as done in the previous sections of this chapter, show some examples from our computer programs, which are able to consider a large number of effects known hitherto (namely, before the first solar-sail mission) for getting a realistic trajectory of high-speed sailcraft.

Although the original description of high-speed sailcraft trajectories was quite different from what we are discussing, it is very useful to begin by considering an earlier observation. For simplicity, let us begin by considering two-dimensional flight: a sailcraft starts its sail-powered flight from a circular orbit about the Sun. Suppose that the maximum lightness number is in the open interval (1/2, 1). If there were only the radial number nonvanishing, then we know from our earlier discussion that the sailcraft trajectory would be hyperbolic, but with decreasing speed. Let us consider a specific, but meaningful example: $l_x = 0.725$. This sailcraft, starting from Earth's orbit, would have a speed of almost 20.1 km/s at 100 AU from the Sun (a speed higher than the Voyager 1 cruise speed). Now, let us think of a different sail control strategy for the same sailcraft: the sail is tilted such that $l_x = 0.534$ and $l_y = -0.242$ (this is possible for a realistic aluminized sail with σ=2 g/m^2, $\alpha = -25.9$ degrees and a sail mean roughness of 10 nanometers, according to Equation 17.3 Figures 16.4 and 16.5). What happens? Because the radial number is higher than ½, the sailcraft first moves outward from the initial circular orbit (as above), but now both angular momentum and energy progressively decrease (statements 1 and 3 in the earlier list) because the transversal number is negative and sufficiently high. Geometrically, the first effect entails that the angle between the sailcraft's position and velocity vectors increases gradually, whereas the second effect implies a speed decrease more quickly than in a pure hyperbolic orbit.

Eventually, a point P^* in space *may* be reached at time t^*, where vector position and velocity, both different from zero, are anti-parallel, namely, the orbital angular momentum **H** vanishes; in addition, in P^* the orbital energy achieves its absolute minimum in the flight. Why "*may*"? Because there are two special options: (1) immediately before P^*—in principle, in the infinitesimal interval (t^*-dt, t*)—the sailcraft accomplishes an impulsive attitude maneuver to get the sail's opposite azimuth, +25.9 degrees; and (2) the sail azimuth is kept at -25.9 degrees. The first case entails an obvious acceleration while the motion continues to be direct (or counterclockwise); namely **H** = 0 is not achieved. A seemingly strange situation happens for the second case : the sailcraft accelerates as well! Why? After reaching **H** = 0, **H** reverses, namely the angle from the position vector to the velocity vector becomes greater than 180 degrees, and the motion becomes clockwise. As a result, a negative azimuth in retrograde motion means acceleration, does it not? Thus, in both cases, the sailcraft accelerates toward the Sun and can actually fly by the star along different paths. However, our surprises are not finished. It can be shown mathematically that the perihelion in each case is *not* the point of maximum speed (as in either classical or generalized Keplerian orbits). Past the perihelion (at 0.20 AU), the sailcraft continues to accelerate until a maximum speed value (83.8 km/s) is achieved very soon. Subsequently, the speed decreases (but not so much) with another striking feature. Due to the radial number that balances more than a half of the solar gravity (at any distance) and to the transversal number that does not cease accelerating the vehicle, the sailcraft speed exhibits a sort of plateau past the initial circular departure orbit. Differently from parabolic and hyperbolic orbits, one can speak of cruise phase indeed.

The attentive reader could argue, OK, it is a fine and intriguing behavior of (nonlinear) solar sailing; but will we gain anything? Well, we above reported a speed of 20 km/s at 100 AU for an intuitive sailcraft control. Now, accurate calculations show that practically the cruise speed amounts to 70 km/s for any $R>9$ AU. We stress that the sailcraft is the same, only the attitude control strategy is different and is constant in HOF again! As a result, this sailcraft is able to reach 100 AU after 7.9 years from launch, instead of 23.2 years. Figure 17.8 shows this case of motion reversal. In particular, the sail orientation in HIF is shown (remember that it is constant in HOF). The capital letters in the figure label special points in chronological order: M maximum distance from the Sun; P, minimum speed; Q, **H** = 0; S, energy vanishes, namely, the escape condition is met; U, perihelion, and W, maximum speed.

The example discussed above is only one element of the large mission class of fast sailing. Figure 17.9A shows how the trajectory profile changes with the negative transversal number and the radial number higher than ½. Note than the motion reversal happens even for lightness numbers lower than that considered above. Figure 17.9B shows the corresponding hodographs. Of course, the unit circles denote the Earth velocity evolution (here assumed as circular for simplicity). Plots are self-explanatory. In particular, note how each hodograph evolves by reversing the curvature vector. The final speeds in the last two examples are considerably higher than the speed of the departure planet.

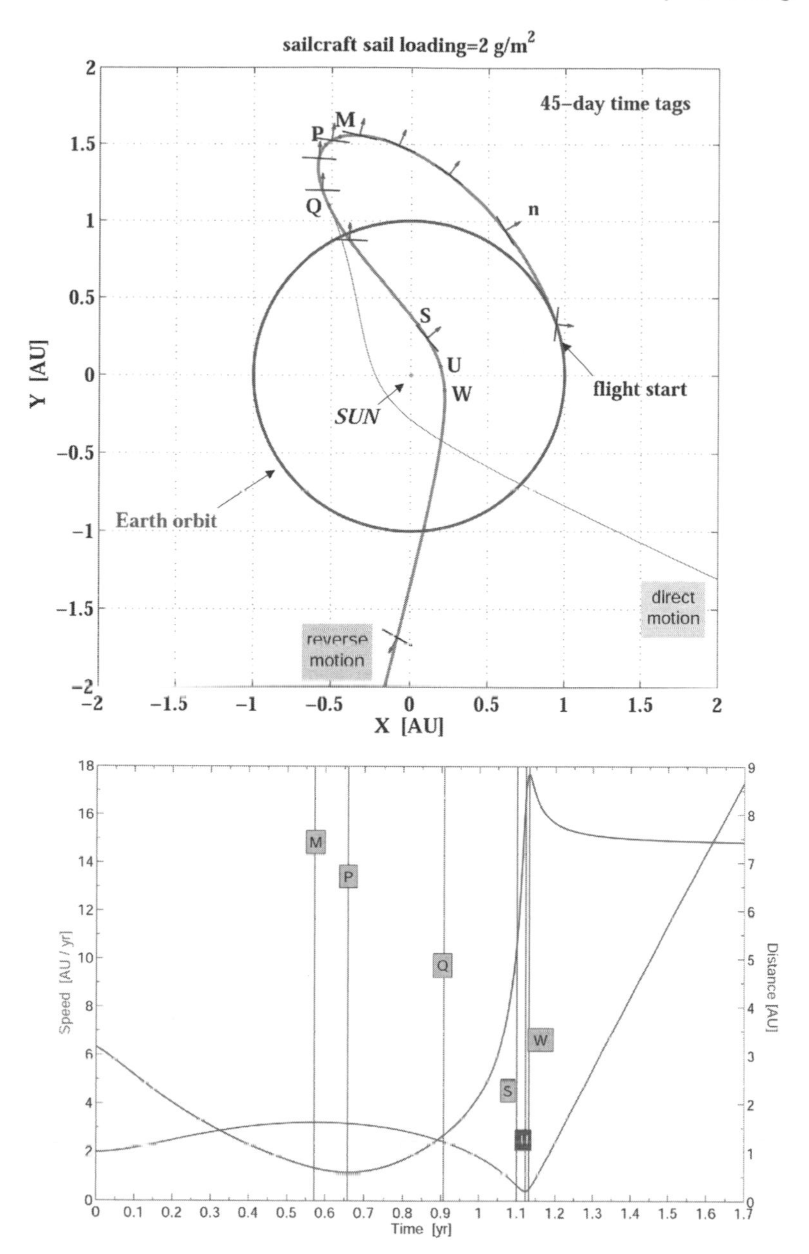

Figure 17.8. Example of *fast solar sailing*. Sun flyby via 2D motion-reversal for escaping the solar system; sailcraft technology is of 2 g/m^2. (A) Pre-perihelion and post-perihelion trajectory arcs with constant-in-HOF sail attitude. The second profile (thinner line), starting from point Q, shows the symmetric direct-motion trajectory. (B) Time behaviors of the Sun–sailcraft distance and sailcraft speed. The time tags are described in the text.

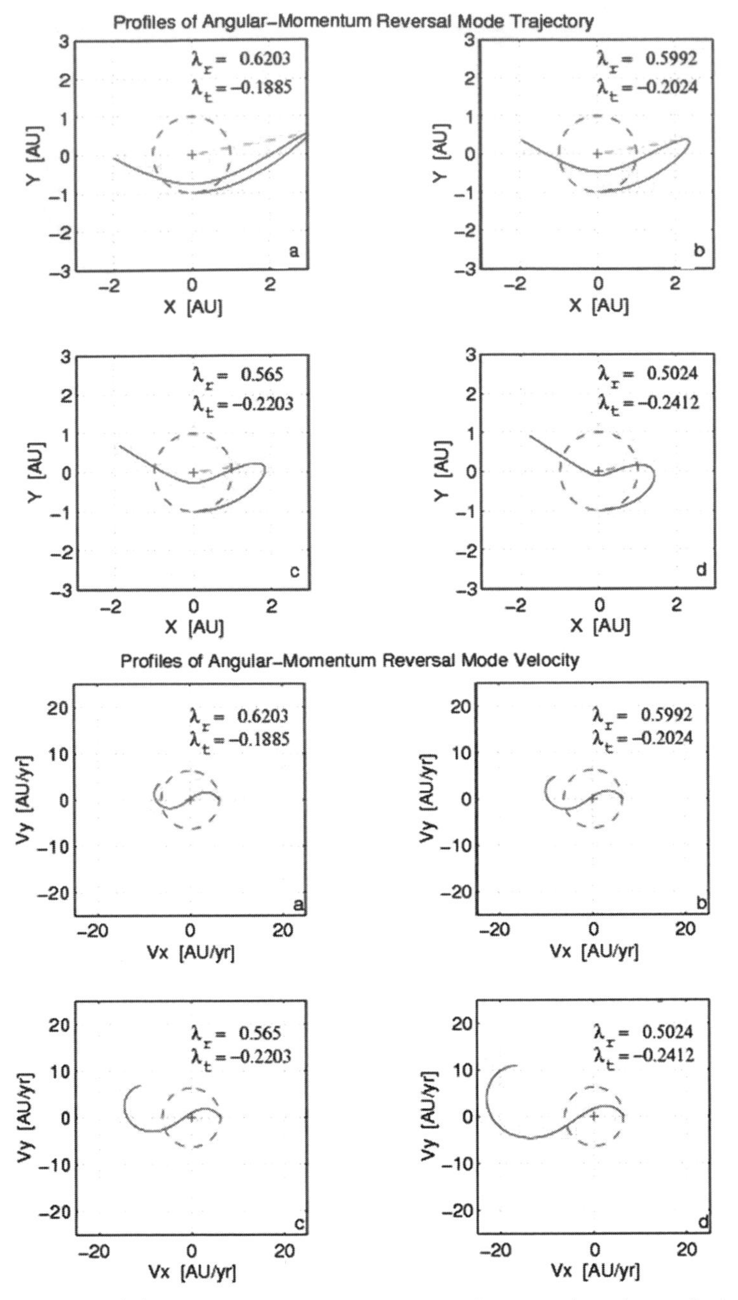

Figure 17.9. (A) Angular-momentum reversal as a function of the negative transversal number and the radial number higher than ½. (B) Hodographs of the trajectories plotted in A. The unit circle denotes the Earth velocity evolution (here assumed to be circular for simplicity). (From G Vulpetti 1997, Further Reading)

Note 2: We have emphasized sailing mode 2. Trajectory of sailing mode 1 is different in the second part, namely, from point Q on. However, both cruise speed and transfer time are exactly equal to those ones of mode 2. This is due to the symmetry of such sailing modes. If you draw the line passing through the Sun (at the origin O) and the point Q, then either trajectory arc can be gotten from the other one by a 180-degree rotation about this line. Then, the final path directions are different. However, this is a very good thing. As a point of fact, by shifting the launch date by some months (depending on the lightness number), one finds *two* launch opportunities per year, every year, for rapidly escaping the solar system toward a prefixed direction. Said differently, the dependence on the relative positions of planets to give energy to spacecraft (often through many flybys) will be only a vague memory when light and large solar sails are made and managed in space.

Note 3: The perihelion value is rather sensitive to the sail attitude. Small midcourse attitude maneuvers may be accomplished to satisfy trajectory constraints. After perihelion, some large attitude maneuver could be designed and safely performed to optimize some index of performance, for example, the time to target.

The 2D trajectories discussed hitherto are ideal; no real sailcraft can strictly move in a plane for a number of reasons. At first glance, this may seem plain; in fact, there are attitude errors, planetary and environmental perturbations, and unmodeled forces. However, there is a nonintuitive cause. If the third component of the lightness vector, which we named the normal number (l_z) in Chapter 16, is different from zero in any finite time interval wherein one requires $\mathbf{H} = 0$, then the trajectory *torsion* diverges as the angle (φ) from $\mathbf{R}$ to $\mathbf{V}$ approaches 180 degrees, no matter how small l_z may be. That brings about a new sailcraft trajectory class (which is three-dimensional and is driven by a sail attitude profile that is not as easy as that above) exhibiting motion reversal. The analysis of such a class involves not only energy and angular momentum, but also trajectory curvature and torsion. It can be proved that the $|\mathbf{H}|$ does not vanish, but passes through a minimum much lower than the angular momentum of the sailcraft departure orbit. After the time of such minimum, the third component of $\mathbf{H}$ reverses and retains its new sign in receding from the Sun.

We would like to end this chapter by discussing a 3D motion-reversal very fast trajectory. The related mission was first presented and discussed at STAIF-2000 (Albuquerque, New Mexico) by author Vulpetti. Sail-system and spacecraft technology has been supposed such that $\sigma = 1.2$ g/m^2, namely, about a factor of 80 better than what envisaged for the first ESA mission. The *maximum* lightness number is 1.21 (equivalent to a characteristic acceleration of 7.18 mm/s^2); that is, the sailcraft could, when necessary, thrust higher than the local gravitational acceleration. Achieving such a goal would not entail the use of nanotechnology. The nominal target of this extrasolar mission is {550 AU, ecliptic longitude = 86.8 degrees, ecliptic latitude = 5.5 degrees}. The distance value means the minimum distance of the Sun's gravitational lens for photons (nothing dealing with gravitational waves); the direction is that from the Galaxy's center to the Sun (the galactic anti-center direction). Figures 17.10 summarizes what may be of concern here. The trajectory arc around the Sun clearly shows that the sailcraft goes above the ecliptic by decelerating, and then it crosses the ecliptic after motion reversal. Perihelion (0.15 AU) is below the ecliptic. The sailcraft reemerges still accelerating because now the optimal sail attitude entails a lightness number of 1.18. The time evolution of sailcraft speed and distance are plotted in Figure 17.10B. The effect of having a sailcraft sail loading sufficiently below the critical density is manifest in the speed behavior: no local maximum takes place. Strictly speaking, if the sail is not jettisoned at some astronomical units, the sailcraft continues to accelerate because the sum of the accelerations is +0.18 $\mu_{\odot}/R^2$ outward. If one jettisons the sail at, for example, 5 AU, the speed loss is 8 percent of the cruise speed. Since a high-technology sail may also work as a multifunction object, there is no compelling reason for jettisoning it inside the solar system; this may be accomplished beyond the heliopause. Note the square-root-like shape of $V(t)$ exhibiting a cruise value of 25.82 AU/year, or 122.4 km/s, or almost three times the escape speed from the solar system at Earth's orbit. According to the criteria set at the beginning of this section, this is a very fast solar-sail mission. The 550-AU target distance is achieved in less than 22 years. A 16-day launch window is found in April, *every year.*

Numerical experiments, regarding Sun flybys via either motion-reversal or motion-direct, have shown that this highly nonlinear effect in solar sailing is possible only if the sailcraft sail loading is lower than 2.20 g/m^2. If medium-term technology exhibits higher values, then *two* direct-motion solar flybys may be used to increase the cruise speed for escaping the solar system. However, one does not get the same performance of the single flyby! The difference between these two modes is significant and depends on the actual σ–value.

Finally, what about the above-mentioned upper bound V^{star}? The superscript *star* means that it depends also on the star that emits light. Is appears obvious that some limit (somewhat less than c) should exist from a couple of evident facts: (1) the Sun has a finite temperature and radius, and (2) the sailcraft sail loading cannot be made arbitrarily small. We suggest some considerations related to dynamics, nanotechnology and space

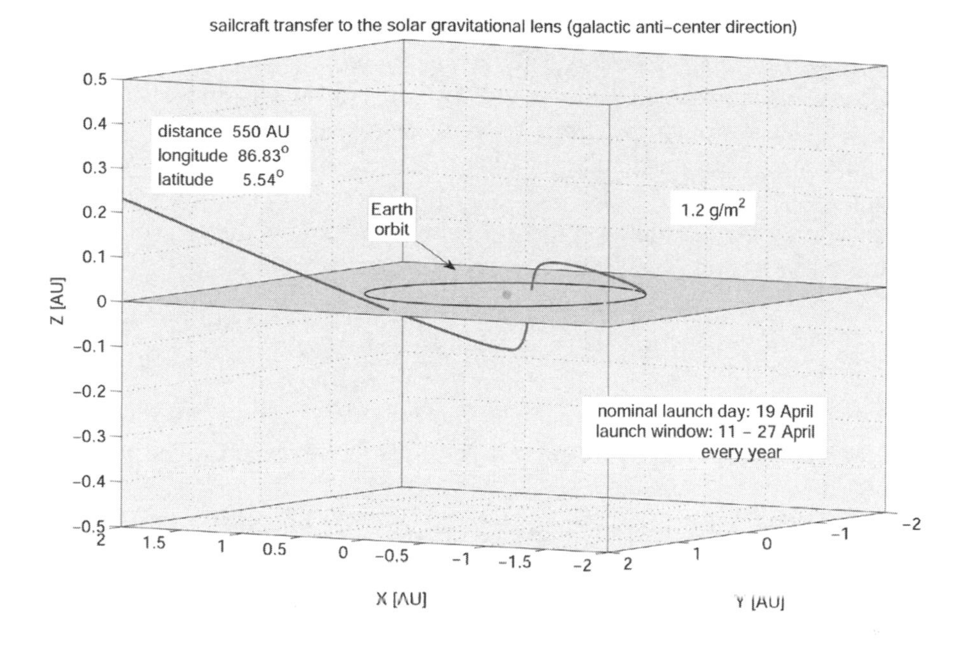

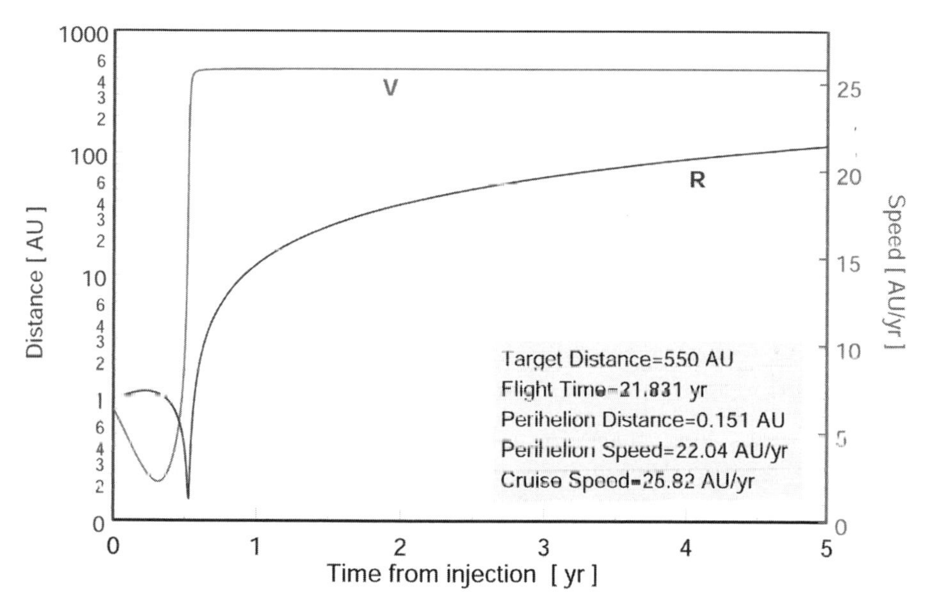

Figure 17.10. Example of 3D *very fast* solar sailing. Sailcraft escapes the solar system via motion-reversal and aims at the target distance of 550 AU. Sailcraft technology has been assumed to be *1.2 g/m²*. (A) Pre-perihelion and post-perihelion trajectory arcs. (B) Time behaviors of the Sun–sailcraft distance and sailcraft speed.

environment (see Chapter 18). As a result, the following very simple expression, $V^{\text{star}} = 2\,\pi\,\sqrt{2\lambda/R_{\text{min}}}$ [AU/yr], would be a good candidate for the maximum realistic achievable speed by an advanced future sailcraft. Projected nanotechnology (from current research) might allow $\lambda \simeq 30$, while the minimum reachable distance (by such a sailcraft) from the Sun might be $R_{min} = 0.05$ AU. Then, $V^{star} \simeq 218$ AU/year $\simeq 1030$ km/s = 0.0034 c. We leave to the reader the several conclusions that can be inferred from such a value in the context of all that has been said in this book.

Further Reading

G. L. Matloff, *Deep-Space Probes*, 2nd ed., Springer-Praxis, Chichester, UK, 2005.

G. L. Matloff, G. Vulpetti, C. Bangs, R. Haggerty, *The Interstellar Probe (ISP): Pre-Perihelion Trajectories and Application of Holography*, NASA/CR-2002-211730, June 2002.

Colin R. McInnes, *Solar Sailing: Technology, Dynamics and Mission Applications*, Springer-Praxis, Chichester, UK, **1999**.

G. Vulpetti, *Overview of Advanced Space Propulsion via Solar Photon Sailing*, lectures held at the Aerospace Engineering School of Rome University, Italy, May 2005, http://www.giovannivulpetti.it.

G. Vulpetti, *Sailcraft at High Speed by Orbital Angular Momentum Reversal*, Acta Astronautica 1997; 40: 733–758.

G. Vulpetti, *3D high-speed escape heliocentric trajectories by all-metallic-sail low-Mass sailcraft*, Acta Astronautica 1996; 39: 161–170.

http://adsabs.harvard.edu/abs/1979Sci...205.1133E.

http://archive.ncsa.uiuc.edu/Cyberia/NumRel/EinsteinTest.html.

Sails in the Space Environment 18

This chapter describes practical and important problems qualitatively.

Gossamer structures like solar sails are very fragile. First-generation solar sails will be manufactured and tested on Earth and, consequently, be required to sustain their own weight in our 1*g* environment. In space they will experience what is perhaps the most hostile environment known to man—space itself. The operating "space" for a solar sail is far more than a mere vacuum, which poses problems in and of itself. Space near the Sun, which is where solar sails will first operate, is bathed in radiation from our star in the form of visible light, ultraviolet photons, x-ray, and gamma rays. The subsequent thermal extremes pose many unique design challenges. The solar wind pummels near-space constantly, and violent storms of charged particles periodically and unpredictably erupt from the Sun and engulf vast regions of interplanetary space much larger than Earth. In general, the inner solar system is not a very friendly place to operate, and it is here that solar sails will be first asked to perform.

Manufacturing: The Environment of Damage and Risk

A solar sail must be lightweight enough to move itself and a payload (in space) when sunlight reflects from it. To meet the design requirements for many of the missions discussed in this book (see Chapters 9 and 17), even the first solar sails must be gossamer-like; hence they will be very fragile. Unfortunately, they must also be large. The sail must be large to reflect enough light to produce thrust and propel itself and its payload to a destination elsewhere in the solar system. First-generation solar sails will have areal densities of 10 g/m^2 or less and be tens of meters in diameter. (This is the loading of the bare sail, not that of the whole sailcraft we denoted by σ in Chapter 16.)

G. Vulpetti et al., *Solar Sails*, DOI: 10.1007/978-0-387-68500-7_18,

At first glance, these sails will resemble common aluminum foil found in many kitchens. Who hasn't attempted to pull aluminum foil off a roll, only to have it hopelessly torn to shreds, forcing you to start over with another piece? However, appearances are misleading. Aluminum foil used in the kitchen is typically 0.013 mm thick, about 10 times thicker than the first-generation solar sails. Now imagine fabricating a sail 100-m by 100-m square out of something ten times thinner than aluminum foil. Not only must the sail be this large, but it has to be strong enough to sustain its own weight under gravity during testing. Even our best materials are too fragile (by themselves) under these conditions and require bracing with cords embedded in them to provide additional strength and to reduce the effects of the inevitable tears. This cord serves the same ripstop function as those found in parachutes. If a tear starts, it will spread until it encounters the cord, where it will be stopped. The edges of the sails are reinforced and securely fastened to the booms during operation. All of these tear-prevention techniques add mass to the sail and must be carefully considered in any sail design.

Launch: Shake, Rattle, Roll, and Outgas

Once the sail is manufactured, it must be folded and stowed for launch. Even though the Cosmos-1 mission by the Planetary Society was unsuccessful in 2005 due to a catastrophic failure in the Russian launcher Volna, the preassembly operations and the assembled spacecraft could give some idea about folding and stowing, as shown in Figure 18.1. In the future, very large solar sails would be folded and stored in small structures within spacecraft for later deployment. Unfortunately, the very factors that make a rocket launch exciting to watch and experience, even vicariously, are the rapid acceleration and the intense vibrations experienced by all things onboard. This vibration environment can damage improperly engineered payloads, shaking them apart before they even make it to space. National space agencies and most, if not all, commercial launch providers require that all payloads demonstrate that they will not shake apart during launch. This requires both analysis and testing during the design and development phases of a project. Again, sails are lightweight and gossamer, making them potentially susceptible to damage from the stresses of launch. Fortunately, with adequate analysis and testing, they can be packaged to survive the launch environment of whatever vehicle they are selected to fly upon.

Another problem during launch is outgassing. A rocket may go from one

Figure 18.1. Pictures of the Cosmos-1 sail folding and packing. (Adapted in color and resolution from the Website of the Planetary Society, U.S., and its related links active in May 2005)

atmosphere of pressure to total vacuum in 8 minutes. A payload riding the rocket experiences the very same pressure change, resulting in a rapid flow of air from the craft to space. A sail has the additional problem of trapped air between folded layers. If the folding is not performed with care, then air bubbles between some layers will form and rush out from between others. The results might range from an inability to deploy (from a bloated sail) to outright destruction (from the rapid out-rush of air, causing a tear). For this reason, testing, called "ascent venting," is performed to simulate the launch environment. It has been shown that sails can be packaged to survive the rigors of launch into space.

Low Earth Orbit: "No-Man's-Land" for Solar Sails

We discussed how easily sails may be damaged during the manufacturing process and during launch. But what about in space? While they are optimized to operate in the low-gravity vacuum of space, one must realize that this environment is neither empty nor benign. The environment of low earth orbit (LEO) is particularly challenging—so challenging, in fact, that it is likely that solar sails will never operate there.

First, there is not a discrete upper boundary to Earth's atmosphere. While the pressures are very low, often lower than many vacuum chambers on Earth, within LEO they are not zero. Broadly speaking, LEO is a region beginning at approximately 160 km altitude extending outward to about 600 km. Within this region, there is a diffuse gas of charged particles, or plasma, formed when sunlight interacts with very high altitude atmospheric gases, giving them enough energy to become ionized and to escape further from Earth. In addition to the plasma, there is a rather significant population of neutral (nonionized) atoms as well. The characteristics of the plasma and neutral atoms are fairly well known and their effects are frequently encountered and measured.

The International Space Station operates well within this region of atmospheric plasma. The net effect is that the station interacts at nearly 8 km/s with the plasma, resulting in an overall drag force on the station, acting to slow it down and drop its orbital altitude. Without frequent propulsive reboost, the Space Station would spiral ever deeper into the atmosphere until it finally burns up and falls to the ground. But the station has been designed to reboost periodically, maintaining its orbit.

The ballistic coefficient is a measure of a spacecraft's ability to overcome air (or plasma) resistance in flight. The ballistic coefficient can be calculated for a body based on its overall mass and surface area. The larger and lighter-weight the spacecraft, the more air resistance it experiences in flight. The smaller or heavier the object, the better it performs. Solar sails are both lightweight and very large, and hence have a very "bad" ballistic coefficient. When unfurled in LEO, the drag on the sail produced by its flight through this residual atmosphere can be very high; larger in magnitude than the thrust the sail experiences by reflecting sunlight. Simply put, a sail flown in LEO will very quickly lose energy by interacting with the ionosphere (despite the fact that it is getting accelerated by reflected sunlight), and find itself on a reentry trajectory. It is easy to compute that a sail shall operate beyond 700 km (nominally); if one takes the upper-atmosphere changes into account, the previous lower limit increases to 750 to 770 km.

Not only does the plasma of LEO put too much drag on the sail, but it also (potentially) causes damage to the sail itself. One of the constituents of the plasma is monatomic oxygen produced when ultraviolet light ionizes a normal diatomic oxygen atom. This monatomic oxygen quickly erodes away many of the materials commonly used in solar sails and other space systems. While not insurmountable, it is still an issue that must be addressed. Recent environmental testing of proposed solar sail materials resulted in their becoming very fragile and, in some cases, disintegrating under the assault of the monatomic oxygen. The combined effects of excessive atmospheric drag, monatomic oxygen, and solar ultraviolet light make LEO a very poor place for solar sails to operate.

The Inner Solar System: At Home for Solar Sails (But Not a Safe Harbor)

Solar sails operate best in the inner solar system and well away from Earth. Sunlight is plentiful and continuous. Away from planetary gravity, there are few mechanical stresses on their tenuous gossamer structures. Though the extreme effects of LEO are not present in interplanetary space, it, too, is far from empty.

Permeating the solar system is a constant stream of small rocky projectiles called micrometeorites. Though micrometeorites are very small, weighing as little as a gram or less in many cases, they are moving very fast. In addition, if they hit the sail, they can potentially damage it. Many of the materials being considered for solar sails were tested under simulated space conditions that included impinging upon them with hypervelocity pellets. Though the sails began to look like Swiss cheese, they remained structurally intact with very little tearing. And since the total reflective area lost from hole formation was very small, there appears to be no impact on the long-term operational performance of the overall solar sail propulsion system.

Ultraviolet (UV) light is a component of the sunlight emitted from the sun. Over time, solar UV light degrades many materials, causing them to become brittle and weak. Some solar sail materials are also affected in this way. Fortunately, even with the increased brittleness from solar UV exposure, tested materials remained intact and functional—even after a simulated exposure equaling several years in the inner solar system.

Close Solar Approaches: Increased Thrust—But at What Cost?

If humans or their robot emissaries are ever to venture to the stars, one of the very few propulsion systems that may ultimately prove feasible is the ultra-thin solar photon sail unfurled as close to the Sun as possible—in a so-called sun-diving maneuver. Metallic monolayer sails tens of nanometers thick satisfy the kinematical requirements of propelling a spacecraft on a millennium-duration voyage to another star. Such sails also seem capable of surviving the thermal environment of a close solar pass, and many of them have tensile strengths equal to the stresses imposed by the consequent high accelerations. Even better solutions may come from nanotechnology, as discussed in Chapter 12.

But alas, that is not the entire story! The near-Sun environment is a far-from-tranquil region. Streams of electrically charged particles—the electrons, protons, and ionized helium nuclei of the solar wind—hurry outward from the Sun at velocities of hundreds of kilometers per second. Although most solar electromagnetic radiation is in the form of relatively benign radio, infrared, or visible light, a considerable fraction is in the ultraviolet, x-ray, or gamma-ray spectral ranges. These photons are energetic enough to ionize sail atoms. As we saw in Chapter 17, considerably better and safer strategies entail solar flybys in either direct or reversal motion. And this may not be a good thing! All this is occurring during a typical "quiet Sun" period. A sun-diving ship foolhardy enough to attempt a close solar pass during the more active phase of the solar cycle would run the risk of encountering the emissions from a solar flare or from the so-called coronal mass ejection (CME, a huge release of the solar-corona matter). Even at Earth's comfortable distance from the Sun, flares can affect weather and disrupt communications. Close up, they would likely be fatal to a sundiving sail.

Solar flares and CMEs are not the same thing, although often they are associated. Perhaps, they might be originated from the same causes, but this is not well understood. From the sailcraft viewpoint, either phenomenon would produce very fast matter that will impinge on the whole vehicle if in the same space-time regions. However, both phenomena appear to be strongly random; therefore, space mission designers are not able to predict them.

When the Sun is quiet, ultraviolet solar photons can knock electrons free from sail atoms. The resulting positive electrical charge on the sail will attract solar-wind electrons to neutralize the ionized atoms. Most electron-atom interactions will be benign, but those involving high-energy electrons will result in sail damage such as reflectivity reduction or degraded mechanical properties.

Mitigation strategies are possible, such as electrically charged grids in front of the sail to moderate electron velocities or layers of protective plastic that evaporate when struck by solar ultraviolet light rather than becoming ionized. But these devices will add mass to the sail and reduce the solar-system escape velocity. Again, nanotechnology could help us in designing solar sails much more resistant to UV.

Actually, it will be far easier to mitigate these effects in ultimate human-occupied interstellar arks than in early robotic interstellar expeditions. To maintain near-Sun accelerations at levels that can be tolerated by human occupants, such craft might require ballast that would be released as the ship accelerates out from perihelion. Charged grids and protective evaporating layers could certainly serve this function as well.

Studies of the interaction between sailcraft and the near-Sun environment are an active field of research. Until NASA or some other space agency launches a probe to survey this region of the solar system, the closest safe solar approach distance will be uncertain. All we can say is that it should be conservatively higher than about 0.1 AU (or 21 solar radii from the Sun). For example, there are further phenomena, related to the *slow* solar wind and not yet known completely, which may affect sails in a manner depending on periods around the minimum of the usual 11-year solar cycle.

State-of-the-Art Materials

The main requirements for solar-sail materials may be summarized as follows: (1) lightweight, (2) strong, (3) highly reflective, (4) easily folded and stored, (5) UV-resistant, and (6) thermally matched to the particular environment in which they will operate.

One support material that meets these requirements is called CP-1. NASA used CP-1, produced by SRS Technologies, Huntsville, Alabama, in its 2005 20-m ground demonstrator program. One of the two 400-square-meter solar sails that NASA tested in hard vacuum conditions was made from CP-1. Smaller samples of it were tested in NASA MSFC's space environmental effects laboratory, where the harsh environment of the inner solar system were re-created. CP-1 performed very well in the tests, and appears to be a promising candidate for first-generation solar sails. We like to stress "first-generation," inasmuch as any plastic support (on which reflective/emissive metals are deposited) forbids the achievement of high lightness numbers.

NASA tested two 400-square-meter solar sails in its ground demonstration program. Instead of CP-1, the second prototype sail used a Mylar sail. Mylar is no stranger to space. It is in use on many spacecraft and significant data exist on its long-term viability in space. While Mylar performed well in the ground demonstrator program, it did not survive well in the deep space environmental effects testing. In fact, researchers report that one of the Mylar samples crumbled when it was removed from the exposure facility. This may not rule it out for use on some solar sail mission applications, but it will certainly not be considered for the broad spectrum of potential missions.

Teonex was also tested in a simulated space environment and is perhaps the most promising candidate identified to date. Teonex samples maintained much of their structural integrity after being exposed to the equivalent of several years' worth of radiation exposure, performing better than either CP-1 or Mylar.

The Japanese flew two space tests of candidate solar sail materials. Pictured in Figure 18.2 is a 2004 test of a solar sail deployed from an S-310-34 sounding rocket. Two types of membrane structures (referred to as clover type and fan type) made by a film of polyimide (which is a long-lasting polymer containing the so-called *imide* monomers, utilized in the electronics industry), were launched with a sounding rocket and deployed sequentially. They were deployed dynamically (i.e., by rotation) in that mission, but some mechanism to deploy membranes statically is required for deploying large membranes. As a point of fact, in August 2006, a membrane of 20 m in diameter was deployed statically in flight using a flying balloon. This is an important step toward the in-orbit deployment demonstration.

The ESA is planning three solar sail missions, the first one (named Geosail) being a technology demonstration mission with a high additional scientific value. At the time of this writing, the envisaged sail materials are CP-1, aluminium, and chromium. Industrial work, paid for by ESA, is in

Figure 18.2. Sail deployment test by the Japan Aerospace Exploration Agency (JAXA) in 2004. (Courtesy of JAXA)

progress aiming at identifying appropriate sail materials, their characteristics, and their behaviors as basic components of the sail system.

Next-Generation Materials Needs

To enable the most ambitious solar sail missions, materials that are lighter, stronger, and more radiation tolerant than the state-of-the-art are required. The overall areal density of the sail material needs to approach or exceed 1 gram per square meter while being strong enough to sustain launch loads and to be manufactured under Earth's gravity. Promising materials with properties approaching these requirements do exist. Carbon composites have many of the desired properties and some promising samples have already been made and undergone some testing. Pictured in Figure 18.3 is a sample of a carbon composite substrate that shows promise for future mission applications.

In Italy, a very preliminary research started in winter 2007 regarding really ultralight and ultra-resistant reflective membranes consisting of doped multi-wall carbon nanotubes (Chapter 12). In addition, such material seems to be transparent to microwaves, thus favoring the design of communication systems onboard sailcraft.

Figure 18.3. Author Johnson shows a very light carbon-composite model of sailcraft. (Courtesy of NASA)

Summary

Solar sails stress our current state-of-the-art materials capabilities, but the needs of first-generation sail missions can now be met. Materials that are manufacturable in large sizes, yet lightweight enough to provide thrust under photon bombardment exist and have been tested in simulated space conditions. The radiation tolerance of candidate materials has been measured, with some outperforming others. Several materials appear to be both foldable and storable with minimal, if any, subsequent deployment issues. From a materials point of view, first-generation solar sails are ready to fly!

Further Reading

NASA/CR-2002-211730, Chapter 4 by author Vulpetti, where there is an introductory mathematical treatment of the sail degradation problem.

Roman Ya. Kezerashvili and Gregory L. Matloff, Solar radiation and the beryllium hollow-body sail: 1. The ionization and disintegration effects, JBIS 2007;60:169–179 (a more comprehensive treatment of the near-Sun environmental issues).

Glossary

Ablation: high-speed evaporation of particles from a heated surface.

Acceleration: the ratio between the velocity change of a body and the time interval during which such variation takes place.

Aeroassist: application of atmospheric drag to perform a space maneuver.

Aerobraking: use of an atmosphere to gradually decrease the energy of a spacecraft orbit.

Aerocapture: orbital capture of a spacecraft by a planet after a single atmospheric pass, namely, a single-pass aerobraking.

Aeroshell: a rigid, heat-resistant structure used to protect a spacecraft during aerocapture.

Allotrope: one of many forms in which some chemical elements take place. Each form differs in physical properties, even though atoms and states of matter (solid, liquid, gas) are the same. Well-known examples are (1) diamonds and coal as forms of carbon, (2) white, red, and black phosphorus, (3) dioxygen (colorless), trioxygen or ozone (blue), tetraoxygen (red).

Antimatter: form of matter with some of its properties reversed with respect to everyday normal matter. Particles of antimatter have the same mass and lifetime as the corresponding normal-matter particles, but all other properties opposite. There also exist (neutral) antiparticles corresponding to neutral particles like neutrons. When normal matter and corresponding particles of antimatter come sufficiently close to each other, they annihilated; that is, their interaction results in practically total conversion of mass to energy. According to the Standard Model (the set of the accepted fundamental theories for physics), matter and antimatter are *specular* to each other. However, in the very last years, there is strong experimental evidence—by analyzing data from the decays of particle B and its antiparticle—that it could not be so. Perhaps this is the first step in ascertaining that asymmetry between matter and antimatter exists, explaining the absence of antimatter in the Universe.

Antimatter rocket: a rocket propelled by the conversion of matter and antimatter fuel into energy.

Apoapsis (or apofocus or apocenter): the farthest an orbiting object (either natural or artifical) gets from its primary attraction body (e.g., the Sun, a planet, etc.).

Aphelion: the farthest a Sun-orbiting object gets from the Sun.

Astrodynamics: the study of the motion of artificial objects in space. In contrast to the celestial mechanics, propulsion is given a central role in astrodynamics in every phase of the space mission (unless it is intentionally excluded during the operational phase of some geodetic satellites, for instance). Astrodynamics has two major branches: trajectory (or orbital) dynamics and attitude dynamics. The former is concerned with the motion of the spacecraft's center of mass (i.e., the translational motion), whereas the latter addresses the motion of the spacecraft *about* its center of mass (i.e., the rotational motion).

Astronomical unit (AU): the radius of a circular orbit where an object of negligible mass would revolve about the Sun in 365.2569 days, according to the two-body Newton laws. 1 AU = 149,597,871 km, approximately the mean distance between Sun and Earth.

Attitude: the orientation of a body in the three-dimensional space (see Chapter 11).

Attitude control system: the hardware and software for controlling, stabilizing, determining, and predicting the attitude of a space vehicle.

Aurora: "light show" in Earth's upper atmosphere associated with impacting solar particles.

Ballute: a cross between a parachute and balloon utilized during aerocapture.

Centrifugal acceleration: one of the accelerations that arise in a rotating system. It is sensed by any particle belonging to a rotating body (see Chapter 11).

Centripetal acceleration: causes any rectilinear path to become curved. It is a pure kinematical concept (see Chapter 11).

Conceptual or thought experiment: an imagined experiment—with no real apparatus—that is used for analyzing what should be observed according to a certain physical theory. It is not a mathematical theorem. Conceptual experiments are very useful in research; they were used fruitfully by Einstein and other famous scientists in the 20th century.

Desorption: evaporation of atoms from a surface caused by some impinging photon beam.

Dynamics: the study of the motion of objects by including the causes that affect the motion.

Ecliptic: as seen from Earth, namely, on the celestial sphere, the *mean* motion of the Sun over the year follows a *great circle*, named the *ecliptic*. The plane of such circle corresponds to the *mean* plane of the Earth's annual path about the Sun. Thus, the term *ecliptic* can be used in the place of *ecliptic plane*. One should not confuse the ecliptic with the mean Earth orbit, which is elliptic and continuously perturbed by planets. To be more precise, since Earth also revolves about the Earth–Moon barycenter, is this point that moves elliptically about the Sun. Most planets go in orbits, about the Sun, close to the ecliptic. The term *ecliptic* stems from being the place where solar and lunar eclipses occur (the ancient astronomers were aware of them).

Exhaust velocity: the exit velocity of expended fuel from a rocket-engine, relative to the rocket vehicle.

Force: a cause inducing velocity changes to a body; in nonrelativistic dynamics, it equals the product of the body's mass and acceleration.

Fullerene: the third allotrope of Carbon (see Chapter 12).

Gravity assist: alteration of a spacecraft trajectory by interaction with a celestial body's gravitational field.

Gravity gradient: a finite-size body, in a nonuniform gravitational field, generally experiences a gravity torque about its center of mass. In space, gravitational fields are not uniform and can affect, via their gravity gradients, the orientation of other bodies, from a man-made satellite to the Moon.

Heliopause: the ideal boundary surface between the solar wind and the interstellar wind.

Inertia: a generic term denoting the aspect of matter that resists change in motion.

Inertial fusion: a nuclear fusion technique using electron beams or lasers to heat and compress the fusion fuel.

Interstellar ark: a human-occupied spacecraft requiring centuries or longer to completing its interstellar journey.

Interstellar ramjet: a concept of a space vehicle collecting interstellar matter as nuclear-fusion fuel.

Ion: an electrically charged atom, namely, a normal atom to or from which electrons have been either added or stripped.

Ion scoop: an electromagnetic device conceived for collecting electrically charged particles in space.

Ionosphere: the layer of the atmosphere that is ionized by the solar photons.

Isotope: two isotopes of the same element have identical numbers of electrons and protons, but different masses since the number of neutrons differ.

Kinematics: the study of the motion of objects without being concerned with the motion causes.

Lagrange (or libration) points: a set of points (stable and unstable) of gravity and centrifugal acceleration equilibrium in the *general* two-body rotating system (e.g. Earth and Moon, Sun and Earth). There exists no equilibrium point in a *restricted* two-body system (e.g., Earth and spacecraft)

Laser: acronym for light amplification by stimulated emission of radiation, a device projecting a coherent, collimated, monochromatic electromagnetic energy beam, usually a visible-light beam.

Lightness number: the magnitude or length of the lightness vector.

Lightness vector: the solar-pressure thrust acceleration vector resolved in the heliocentric orbit frame, centered on the sailcraft, and taking the local solar gravitational acceleration as the normalization factor. Its properties are discussed in Chapter 16.

Magnetic sail or Magsail: a sail concept for slowing an interstellar spacecraft by the electromagnetic reflection of interstellar ions.

Magnetosphere: the (large) volume around Earth where its magnetic field is compressed and bounded by the solar wind.

Maser: a laser operating in the microwave region of the electromagnetic spectrum.

Mass ratio: the ratio of a rocket's mass prior to ignition (including fuel) to its mass at burnout.

Mini-magnetosphere or M2P2: a concept of magnetic space-propulsion device operating by the reflection of the solar-wind ions.

Momentum: for a massive body, *linear* momentum is the product of its mass by its velocity. For the particles of light (photons), it is the ratio of its energy to the speed of light. Momentum is a physical quantity.

Nanometer (nm): one billionth of a meter.

Nanophysics: the branch of physics dealing with the nonclassical phenomena exhibited by either single-atom or many-atom aggregates of 0.1 to 100 nanometers in size; the lower range of these systems is dominated by the laws of quantum mechanics (which even holds down to the atomic nucleus and elementary particle levels, a millionth of a nanometer or shorter!).

Nanotechnology: technology at the molecular and atomic range (from 1 nm to 100 nm, typically); such technology will allow making and utilizing devices and structures as systems having novel physical and chemical properties due to their small sizes.

NEO: Near Earth Object, an asteroid or comet orbiting close to the Earth–Moon system.

Nuclear fission: a nuclear reaction in which most energy is released as kinetic energy of heavy nuclei split to produce lower-mass "daughter" nuclei.

Nuclear fusion: a nuclear reaction in which low-mass atomic nuclei combine to produce more massive particles, but with energy release.

Oort comet cloud: a reservoir of some trillion comets reaching perhaps halfway to our Sun's nearest stellar neighbors.

Periapsis (or perifocus or pericenter): the closest an orbiting object (either natural or artificial) gets to its primary attraction body (e.g., the Sun, a planet, etc.).

Perihelion: the closest a Sun-orbiting body gets to the Sun.

Planck constant: any type of light (the rainbow colors, the oven microwaves, the solar ultraviolet and radiology x-rays, etc.) appears in the form of noncontinuous pieces (quanta) of energy. The energy of a particle of light (photon) is equal to the product of its vibration frequency (ν) by a universal constant (h), called the Planck constant. In SI units, its value is $6.6260693 \times 10^{-34}$ J s (joule times second). The energy of a photon can be expressed by $E = h\nu = hc/\lambda$, where λ denotes the wavelength (see *speed of light*).

Plasma: the fourth state of matter, typically any ionized gas. There, atoms are stripped of some or all their electrons; however, such atoms cohabit with the electrons and form a conductive, though macroscopically neutral, gas.

Pole sitter: a concept of spacecraft permanently situated in a high-latitude region of the celestial sphere.

Pressure: given a force of magnitude F acting perpendicularly to a surface of area A, pressure is defined as the ratio F/A. In the international units (metric) system (called the SI units), pressure is measured in pascal (Pa), which is the force of 1 newton (N, approximately 102 grams) pushing on a surface of 1 square meter. In weather forecasting, the usual unit is the hPa (hecto-pascal or 100 Pa). In the U.S. system of measurements, 1 PSI (pounds/square inch) = 6895 hPa.

Radiometer: a device for measuring the energy of light that crosses a unit surface in a unit time interval. Normally, this instrument is used for visible and infrared light, but it can be employed in other regions of the so-called electromagnetic spectrum.

Ram-augmented interstellar rocket (RAIR): a concept of spacecraft collecting interstellar ions as supplemental reaction mass.

Ripstops: a network of strengthening fibers embedded in a sail film to reduce the severity of rips and tears.

Sailcraft: a space vehicle endowed with a sail that functions as momentum exchange; it acts as a propulsive device receiving momentum from an *external* source. Therefore, a sailcraft is quite different from a rocket spacecraft. Chapter 7 discusses sailcraft and their new features with respect to a rocket. Usually, sailcraft = sail-system + spacecraft, the latter term regarding all systems different from the sail assembly (which includes the structures shaping the bare sail). Spacecraft and sail system (or assembly) are physically connected.

Sailcraft (sail) loading: the ratio between the sailcraft mass and the effective sail area, usually expressed in grams per square meters. It is a basic parameter in sailcraft dynamics.

Solar constant: the Sun emits a flux of photons, the energy of which ranges from radio wave to X-rays and gamma-rays. The total energy that flows through 1 square meter, at rest and perpendicular to the incoming solar photons at 1 AU, is known as the *solar constant*, say Φ. Its technical name is the *total solar irradiance* (TSI) at 1 AU. In the last two decades, researchers have discovered—via satellites equipped with special radiometers—that TSI is *not* constant. Its variability—though slight—reflects a number of Sun-related phenomena, some of which have not yet been well understood. The current accepted *mean* value of TSI at 1 AU amounts to $\Phi = 1366.1$ watts/square meter, which corresponds to a pressure of light $P = \Phi/c = 4.557 \cdot 10^{-6}\ Pa$; it is a very small value compared to everyday standards, but not as small in space as it may seem. Because solar light expand spherically into space, this photon pressure scales as $1/R^2$, where R denotes the distance (expressed in AU) between the Sun and a space body. For instance, at the mean distance of Mars, $P = 1.972 \cdot 10^{-6}\ Pa$, whereas at 0.2 AU from the Sun one gets $P = 113.9 \cdot 10^{-6}\ Pa$. Such variability can be utilized for navigating in space by appropriate sails.

Solar flare: an explosive emission of plasma and electromagnetic radiation (photons) from the Sun's surface.

Solar wind: the Sun emits a very high number of fast massive particles, essentially protons and electrons (95 percent), alpha-particles (or nuclei of helium, 4 percent), and other ionized atoms. Such particles form what is known as the *solar wind*. (Evanescent particles that physicists call neutrinos are emitted as well, but they cannot be utilized for space propulsion). Solar wind should *not* be confused with the solar photon flux, which shall be utilized for solar sailing.

Speed of light: in a vacuum, light propagates with a constant speed, normally indicated by c. It is equal to 299,792,458 m/s, often shortened by 300,000 km/s. Light behaves also as waves; they exhibit wavelengths (usually denoted by λ), which is the space scale where the electric and magnetic fields oscillate by completing one cycle. The number of cycles completed in one unit time is named the *frequency*, say ν. One has the basic relationship $c = \lambda \, \nu$. (For instance, a radio wave 300 meters long vibrates about one million times every second, whereas the yellow light vibrates about 500,000 billion times per second). This relationship holds for any type of waves; however, one has to be careful in using the correct speed of wave propagation. For instance, if one deals with the sound waves in air, one has to replace c by the usual speed of sound (343 m/s, approximately in dry air at 20°C or 68°F).

Sun diver: a maneuver type used by a sailcraft to approach the Sun as closely as possible.

Tether: a long cable that can be used in space for orbit modification. Momentum-exchange tethers are mechanical devices. Electrodynamic tethers interact with the planet's magnetic field, if any.

Thermodynamics: the branch of physics that studies macroscopic real systems from the viewpoint of their energy exchange (in particular as heat), temperature, pressure, volume, and so forth. Thermodynamics is fundamental also in designing practical working devices like refrigerators, air-forced circulation systems, car motors, ship and aircraft engines, space rocket engines, and so on. The 19th century saw the development of thermodynamics as a modern science that allowed inventing and designing basic transportation systems such as trains and steamships.

Thrust: the force from any propulsive device; in particular, for a rocket, this force comes from exhausting the fuel.

Technology readiness: a NASA-developed system to track the developmental status of a space propulsion system.

World ship: an interstellar ark large enough so that its habitable interior approximates the terrestrial environment.

Wrinkles: elastic (i.e., recoverable) sinusoid-like undulations of the sail membrane due to compressive forces; wrinkles should not be confused with *creases*, which are inelastic deformations, especially when a membrane is coated by metal films. Sail folding and handling can cause different-pattern creases.

Index